Education, the Environment and Sustainability

Edited by
Kai Horsthemke

Routledge
Taylor & Francis Group
LONDON AND NEW YORK

First published 2024
by Routledge
4 Park Square, Milton Park, Abingdon, Oxon OX14 4RN

and by Routledge
605 Third Avenue, New York, NY 10158

Routledge is an imprint of the Taylor & Francis Group, an informa business

British Library Cataloguing in Publication Data
A catalogue record for this book is available from the British Library

ISBN13: 978-1-032-68468-0 (hbk)
ISBN13: 978-1-032-68471-0 (pbk)
ISBN13: 978-1-032-68473-4 (ebk)

DOI: 10.4324/9781032684734

Typeset in Myriad Pro
by Newgen Publishing UK

Publisher's Note
The publisher accepts responsibility for any inconsistencies that may have arisen during the conversion of this book from journal articles to book chapters, namely the inclusion of journal terminology.

Contents

Citation Information

The following chapters were originally published in the journal *Ethics and Education*, volume 16, issue 2 (2021). When citing this material, please use the original page numbering for each article, as follows:

Chapter 1

The task of education as we confront the potential for social and ecological collapse
Vanessa De Oliveira Andreotti
Ethics and Education, volume 16, issue 2 (2021), pp. 143–158

Chapter 2

'Landing on Earth:' an educational project for the present. A response to Vanessa Andreotti
Sharon Todd
Ethics and Education, volume 16, issue 2 (2021), pp. 159–163

Chapter 3

Education, anthropocentrism, and interspecies sustainability: confronting institutional anxieties in omnicidal times
Helena Pedersen
Ethics and Education, volume 16, issue 2 (2021), pp. 164–177

Chapter 4

Animal advocacy, fear and loathing in academia: a response to Helena Pedersen
Kai Horsthemke
Ethics and Education, volume 16, issue 2 (2021), pp. 178–181

Chapter 5

Be the village: exploring the ethics of having children
David Chang
Ethics and Education, volume 16, issue 2 (2021), pp. 182–195

Chapter 6
Got milk? from growing strong bones to nurturing idealized subjectivities
Samantha Deane and Annie Schultz
Ethics and Education, volume 16, issue 2 (2021), pp. 196–208

Chapter 7
Psychoanalytic ecofeminist Dorothy Dinnerstein: theorizing the roots of rapacity
Gregory Bynum
Ethics and Education, volume 16, issue 2 (2021), pp. 209–221

Chapter 8
Educating in and for uncertainty. climate science, human evolution and the legacy of Arne Naess as guidance for ecological practice
Margarita García-Notario
Ethics and Education, volume 16, issue 2 (2021), pp. 222–235

Chapter 9
Co-Creation in the Commonwealth: Understanding Right Relationship in Place
Mark Beatham
Ethics and Education, volume 16, issue 2 (2021), pp. 236–248

Chapter 10
Youth power—youth movements: myth, activism, and democracy
Lynda Stone
Ethics and Education, volume 16, issue 2 (2021), pp. 249–261

Chapter 11
White, Green futures
Cortland Gilliam
Ethics and Education, volume 16, issue 2 (2021), pp. 262–275

Chapter 12
Spiritual exercises in times of climate change
Daniel P Gibboney
Ethics and Education, volume 16, issue 2 (2021), pp. 276–287

Notes on Contributors

Mark Beatham is Associate Professor of education at SUNY College at Plattsburgh in Plattsburgh, New York, USA. He teaches courses in the history, politics, philosophy, and psychology of education as well as in the Honors Center, where he is on the Executive Board. His principal research area is philosophy of education.

Gregory Bynum is Associate Professor in the School of Education at the State University of New York at New Paltz, USA. His teaching and research interests include gender equality and education, anti-racist education, and respect for childhood intellect. He enjoys playing the recorder.

David Chang is a teacher educator and counsellor, and Ph.D. candidate in philosophy of education at Simon Fraser University, Canada. A student of the Zen Buddhist tradition, his main research interests are contemplative practices for environmental education and lifestyles that support ecological well-being.

Samantha Deane, Ph.D., directs the project for Formative Leadership Education at Boston College, Massachusetts, USA. A philosopher of education, her research focuses on agency and democratic education. She is the author of the forthcoming book, *Democratic Education in an Armed Society: Learning to Live with Guns.*

Vanessa De Oliveira Andreotti started her appointment as Dean of the Faculty of Education of the University of Victoria in July 2023. She is a former Canada Research Chair in race, inequalities and global change and a former David Lam Chair in multicultural education. She is the author of *Hospicing Modernity: Facing Humanity's Wrongs and the Implications for Social Activism* (2021) and one of the co-founders of the Gesturing Towards Decolonial Futures (GTDF) Arts/Research Collective.

Margarita García-Notario teaches courses at SUNY Plattsburgh in Spanish, philosophy, mythology, ancient cultures, and deep ecology. Her doctorate in philosophy of education is from Universidad Complutense (Madrid), and her degrees in philosophy and theology from Universidad de Navarra (Pamplona). Among her publications is *Filosofía Profunda y Educación*.

Daniel P. Gibboney recently graduated from the University of North Carolina at Chapel Hill's School of Education, USA. His research and teaching interests are in the history of science, youth studies, and Michel Foucault. He is currently completing a monograph on the history of childhood and developmental psychology in early twentieth-century America.

Cortland Gilliam is a poet, educator, cultural organizer, and Ph.D. candidate in the School of Education at the University of North Carolina at Chapel Hill, USA. His academic work parses cultures of school discipline, civics education, and histories of Black youth contributions to liberation movements of the late twentieth century.

Kai Horsthemke is Visiting Professor at the University of the Witwatersrand, South Africa, and at KU Eichstätt-Ingolstadt, Germany, and a fellow at the Oxford Centre for Animal Ethics. His teaching and research interests include animal ethics, environmental education, African philosophy, indigenous knowledge systems, and social epistemology.

Helena Pedersen is Associate Professor in education at the University of Gothenburg, Sweden. She is the author of *Schizoanalysis and Animal Science Education* (2019) and *Animals in Schools* (2010). She is on the Editorial Board of the Critical Animal Studies book series and co-founder of University of Gothenburg's Network for Critical Animal Studies in the Anthropocene (GU-CAS).

Annie Schultz is Assistant Professor of classical and liberal education at Flagler College in Saint Augustine, Florida, USA. Her research interests span a variety of areas in philosophy of education, including environmental education, aesthetics, art education, and gender studies.

Lynda Stone is the Samuel M. Holton Distinguished Professor Emerita at the University of North Carolina at Chapel Hill, USA. Her current scholarly endeavors with particular interests in democratic societal and school reform include a work in progress entitled *Challenging Certainty: Essays in Ethics, Politics, and Schooling*. Honors include presidencies of the John Dewey Society and the American Educational Studies Association. She received the Lifetime Achievement Award in 2018 from the Dewey Society.

Sharon Todd is Professor of education at Maynooth University, Ireland. She has published widely on embodiment, social justice and ethics in education and is currently focusing on aesthetic responses to climate emergency. Her most recent book is *The Touch of the Present: Educational Encounters, Aesthetics, and the Politics of the Senses* (2023).

Introduction

Kai Horsthemke

ABSTRACT

The 17th Biennial Meeting of the International Network of Philosophers of Education (INPE) was scheduled to take place from 28 to 31 July 2020 at the Universidad Nacional Autónoma de México, Mexico City. Of course, there is something ironic about convening a conference on the environment and sustainability that would require presenters to utilize unsustainable modes of transport in order to participate. As it turned out, because of the outbreak and rapid global spread of the new and highly infectious coronavirus, Covid-19, the conference was cancelled and replaced by an online event held on 7 and 8 November 2020, both premieres in the history of INPE. The essays collected in the Special Issue of *Ethics and Education* (16/2) are replicated in this volume, thereby being made available to a wider readership. If anything, these essays are probably even more relevant now than when they were first published, in May 2021. They will hopefully contribute towards the unmasking and undoing of the various kinds of denialism that have held us in its grip and that continue to thwart attempts to establish a sane and morally sustainable set of relationships between us, human beings, and other animals and the animate and inanimate environment.

The world was not given to you by your parents, it was lent to you by your children.
(Kenyan proverb)

'One of the greatest challenges – if not *the* greatest – facing humankind at the beginning of the 21st century is arguably the state of our planet and, coupled with this, our relationship with the natural environment. Most, if not all, other concerns – however significant – are necessarily secondary in this regard. The human impact on the environment has been, and continues to be, enormous. Human population growth and advances in technological ability and control continue producing previously non-existent environmental problems. What is at stake here is nothing less than the survival of the earth in its present state, as being inhabitable, and therefore also human survival itself. At the very least, it is a matter of the quality and conditions of our lives, present and future. If this is correct, it follows that one of the greatest priorities – perhaps *the* greatest – of academic research, scientific, philosophical, educational and other, may be how to arrest and possibly reverse the present decline. There are difficulties, of course. The problem is not only climate change denialism, that is, doubts that the diagnoses and prognoses are correct. Substantial philosophical and pedagogical questions also concern the articulation and definition of the approaches we human beings should adopt in response to the environmental predicament/s. Further questions are raised about the metaphysical and ethical foundations of our concern for our planet, and about educational and pedagogical implications. Should educational policy and practice, for example, be informed by a concern for nature and the environment for our (human) purposes? Or should we teach and learn for the natural environment in and for itself?'

This introductory paragraph is taken from the Call for Papers for the 17th Biennial Meeting of the International Network of Philosophers of Education (INPE) that was scheduled to take place from 28 to 31 July 2020 at the Universidad Nacional Autónoma de México, Mexico City, with Renato Huarte Cuéllar as chair of the site committee. Of course, and this is something that was pointed out by a number of concerned colleagues, there is something ironic about convening a conference on the environment and sustainability that would require presenters to utilize unsustainable modes of transport in order to participate. As it turned out, because of the outbreak and rapid global spread of a new coronavirus, Covid-19, the conference was cancelled, a premiere in the history of INPE. All of a sudden, a new challenge was facing humankind at the beginning of the 21st century, as the number of infections and fatalities rose exponentially, countries went into lockdown mode, and economies all but ground to a halt. But is it really a *new* challenge, one that is unrelated to the climate crisis, rapid losses of biological diversity and our relationship with the natural environment? No. At the heart of the health crisis that has engulfed our planet and that has targeted human beings in particular, in the form of a deadly virus, is an obsession with economic growth and human expansion – in terms of sheer numbers, resulting in unsustainable population density, progressive encroachment on natural habitats, and exploitation and destruction of the biosphere.

Seventy-five percent of all infectious diseases are therionotic, that is, they originate in non-human animals (Jones et al. 2008; Belay et. 2017). The conditions of factory farming, intensive animal husbandry and rapid human encroachment on non-human habitats have proved ideal for the development, spread and increasing mutations of killer viruses, which are transmitted to humans via the ingestion of animal products. The so-called Spanish Flu that lasted from January 1918 to December 1920 (so-called because the Spanish press was the only press at the time that presented uncensored accounts of this uncommonly deadly influenza pandemic) appears to have started with the infection of a US poultry farmer and spread following his recruitment for the US armed forces during the first world war. This respiratory virus was harboured in birds, mutated and later migrated to pigs, fatally afflicting humans before long. The SARS (severe acute respiratory syndrome) virus and the Ebola virus, both appearing in the early 21st century, can similarly be traced back to the proximity and consumption of captive domestic and more exotic animals, like civets, raccoon dogs, fruitbats and flying foxes. The new coronavirus is no different. It has been located in bats and pangolins, who have proved ideal hosts – they are carriers but themselves insusceptible to the virus. While fruitbats and flying foxes have been among the 'bush meat' favoured in many parts of the African continent, bats and pangolins have been popular features of the so-called wet markets in parts of Asia, traditional localities where living animals can be purchased and are often slaughtered upon or immediately prior to purchase. The outbreak of a new strain of coronavirus in Danish mink farms is a further case in point, albeit with little to no physical consequences for humans – owing to the rapid 'destruction' of millions of mink following the discovery.

As the spread of the virus started to assume dramatic proportions, I received a meme about a conversation between Homo sapiens and Mother Nature. Homo sapiens protests, 'There's no way we can shut everything down in order to lower emissions, slow climate change and protect the environment.' Mother Nature responds, 'Here's a virus. Practice.' Within a few weeks of the viral outbreak and spread, the smog over many parts of China and northern Italy, following curfews and industrial lockdowns, had cleared visibly – as satellite photos showed. Swans, dolphins and species of fish seldom seen returned to the canals of Venice. Coyotes were spotted in the streets of San Francisco. Wolves are being seen with increasing frequency in Germany, after having been on the brink of extinction. A wild herd of about 120 Kashmiri goats ventured from the Great Orme into Llandudno in Wales. I am not citing these cases in order to spread some ill-founded optimism about the crisis (note the singular) we find ourselves in or about the post-corona situation. I have no intention to trivialize the hardships we have faced and may come to face again in future. Quite the contrary: it is a time for reflection and for lasting humility on our part. Helena Pedersen, one of the invited keynote speakers, has in an earlier essay posed a question that is of fundamental relevance in the present context: 'What happens with education if it acknowledges that the world does not need humans, and is likely to thrive ignorant of human existence?' (Pedersen and Pini 2017, 1053) This tallies with Steven Best's contentions (2014, 119 and 166, respectively) that, while earthworms, dung beetles, butterflies and bees are important to the integrity and diversity of nature, 'human beings could be removed from earth ecosystems with positive effect', and that Homo sapiens is 'the one species the earth could well do without'. This theme is picked up by David Chang in his essay, which begins with the diagnosis of rapid human population growth as a chief contributing factor to the ecological crisis our planet is facing. Chang does not advocate coercive measures for population control but instead favours an approach to environmental education that encourages students to consider the assumptions and implications of having children and that thereby fosters a sense of social and ecological responsibility.

In her keynote address, which was also the Terry McLaughlin memorial lecture, Vanessa De Oliveira Andreotti points out that formal education has been complicit in the reproduction of historical and systemic violence, as well as unsustainability. The task of a substantially revised and re-imagined formal education is *both* to enable mastery of relevant knowledge and skills, that is personal empowerment, *and* to contribute towards 'expansion of our capacity to process collective and personal traumas and to sense, relate and imagine "otherwise"'. The second part of Andreotti's essay presents *Education 2048*, a thought experiment designed to enable uncomfortable but necessary conversations about the role of education in preparing us all for confronting the potential for and, indeed, the increasing possibility of social and ecological collapse during our lifetime. Sharon Todd's empathic response constitutes an initial contribution to these vitally important conversations.

Helena Pedersen's contribution to this volume poses the question: What prevents education from becoming a transformative force in times of 'omnicide', that is, 'the annihilation of everything'? She locates at least part of the response in 'institutional anxiety', which constitutes a (social-) psychological barrier to radical change. In particular, she discusses anxiety related to the moral standing of non-human animals as a threat to human exceptionalism in educational practice and research. (In his response, Kai Horsthemke discusses a recent manifestation of such anxiety at a university in South Africa.) Confronting these anxieties, Pedersen argues, could pave the way for new modes of being and acting within academia and provide interspecies ethics, justice and sustainability with an opportunity to flourish in omnicidal times.

The phenomenon of fear also appears in Margarita García-Notario's essay. She argues that educating in and through fear is bound to be unhelpful in generating constructive and creative responses to the issue of climate change and the general state of our planet. Drawing on the deep ecological approach pioneered by Arne Næss, García-Notario proposes the principles of 'deep questioning, deep experience and deep commitment' as a suitable guide for sustainable educational and ecological practices. Gregory Bynum pays tribute to a contemporary of Næss, environmental activist and feminist psychoanalytical scholar Dorothy Dinnerstein, who proposed that the sharing of childcare responsibilities would have a significant impact on male rapacity and domination, effectively curbing androgenic environmental damage. Her theories, which were not widely endorsed at the time of their publication, receive a sympathetic response from Bynum. He defends them against Dinnerstein's critics and contends that it is not only women and the natural environment that stand to benefit if Dinnerstein's recommendations are taken seriously, but also men themselves.

In their contribution to the discussion of the moral/ethical dimensions and racial and sexual politics of 'the hidden curriculum of what we eat, who we eat with, and the significance afforded this moment of the school day', Samantha Deane and Annie Schultz are concerned with 'the social and political consequences of the naturalization of the care work of women and one important symbolic representation of this work: milk'. They problematize the extent to which both this care work and the functions of non-human females in the dairy-industrial complex are taken for granted on the assumption that they are 'natural'. Deane and Schultz conclude by contending that we should desist from '"milking" the care of women and [other] female bodies' and begin considering 'the ontological significance of the products children consume'. Milk does a body good, *if* it does, only at the expense of other bodies.

Mark Beatham's interest in his contribution is with place-based education, that is, nurturing through (public) education an understanding of shared spaces and proper or right relationships with and within these. Proper relationships, he argues (drawing on the work of Wendell Berry), 'preserve the means of co-creation between human culture and creation': 'Sustainability requires a proper relationship between culture and nature'. Although Beatham does not employ these terms, it is not difficult to see how this kind of education might not only have an emancipatory function but also encourage activism among learners and students, motivating them to 'add to the commonwealth through the accumulation of knowledge, efforts, and commitment'.

Unsurprisingly, young Swedish activist Greta Thunberg and the *Skolstrejk för klimatet/Fridays for Future* movement she started in 2018 are mentioned in several essays (Pedersen, Garcia-Notario, Stone, Gilliam). The three papers by Lynda Stone, Cortland Gilliam and Douglas Gibboney were presented as a symposium during the INPE online conference in November 2020, the topic being 'Together America and the World: Environment Issues, Participation, and Education', each presentation raising a set of different albeit-related issues. Common to all three is the question of participation, democratic and otherwise – by whom, for whom and in what form. Stone is centrally concerned with youth power, youth movements and youth activism in response to the climate crisis. Gilliam, too, discusses youth activism, but he highlights the fact that 'non-White activists' have been sidelined and assigned 'back seats' in a global youth climate movement that has been represented predominantly by 'White and Western climate activists'. Gibboney points out that the climate crisis is actually a crisis of facts that has climate science believers pitted against climate change denialists, advocates of alternative facts and conspiracy theorists. His essay explores activism in the light of 'possibilities of shared practices' in times of 'unshared realities'. Gibboney's 'spiritual exercises', in this regard, complete the arc that began with Andreotti's reference to a need for activating the imagination, for creating new sensibilities (see also Todd, in this volume).

Philosophy in all its forms and traditions has always been centrally preoccupied with truth. However, in recent years truth appears to have become unfashionable. Climate change and coronavirus denialism and conspiracy theories have all contributed to the current state of affairs in which truth and facts are treated with disdain. For example, the ideas that truth is not objective or universal and that there can be 'alternative facts' have taken hold not only of popular culture but also of politics, to the obvious detriment of the latter (Horsthemke 2017). The precedence given to assorted political and ideological (not to mention personal narcissistic) agendas, along with the rising popularity of pseudoscience, has led to a decline of truth both as a serious subject and as an intellectual tradition. Climate change denialism and coronavirus denialism are objectionable not only because they peddle blatant untruths but also because they are cynical and cruel. They demean the many millions, human and non-human, who have forfeited their health and/or their lives as a result of the changing climate and the new pandemic. It is my sincere hope that the essays collected in this volume will contribute towards the unmasking and undoing of the pernicious relativism (cultural, moral and epistemological) that has held us in its grip and that continues to thwart attempts to establish a sane and morally sustainable set of relationships between us, human beings, and other animals and the animate and inanimate environment.

References

Belay, E.D., J.C. Kile, A.J. Hall, C. Barton-Behravesh, M.B. Parsons, S. Salyer and H. Walke. 2017. "Zoonotic Disease Programs for Enhancing Global Health Security". *Emerging Infectious Diseases* 23(13): S65–S70.

Best, S. 2014. *The Politics of Total Liberation: Revolution for the 21st Century* (New York, Palgrave Macmillan).

Horsthemke, K. 2017. "'#FactsMustFall'? – Education in a Post-truth, Post-truthful World". *Ethics and Education* 12(3): 273–288.

Jones, K.E., N.G. Patel, M.A. Levy, A. Storeygard, D. Balk, J.L. Gittleman and P. Daszak. 2008. "Global Trends in Emerging Infectious Diseases". *Nature* 451(7181): 990–993.

Pedersen, H. and B. Pini. 2017. "Educational Epistemologies and Methods in a More-Than-Human World". *Educational Philosophy and Theory* 49(11): 1051–1054.

The task of education as we confront the potential for social and ecological collapse

Vanessa De Oliveira Andreotti

ABSTRACT

This article invites us to consider the task of education as we face the end of the world as we have known it. The first part of the article gives an overview of global and educational challenges, drawing attention to how formal education has been complicit in the reproduction of historical and systemic violence, as well as unsustainability. This section also offers a distinction between educational approaches that focus on personal empowerment and the mastery of knowledge and skills, and educational approaches that see the role of education in association with the non-coercive re-arrangement of desires and with responsibility before will. The second part of the paper presents a psychoanalytic experiment that attempts to create a space and the dispositions necessary for difficult conversations about the role of education in preparing us all to confront the potential for social and ecological collapse in our lifetime.

Introduction

This article is based on a Terence H. McLaughlin lecture presented at the International Network of Philosophers of Education conference on 7 November 2020. It starts with a brief overview of enduring global challenges and what I see to be challenges for the context of education as an academic discipline and professional field. Next, I make a distinction between approaches to education focused on mastery of knowledge, competencies and skills and approaches that focus on depth of intellectual and relational rigour. In the last part of the article I offer a psychoanalytic thought experiment about the role of education in confronting the potential or likelihood of social and ecological collapse in our lifetime.

I would like to acknowledge the fact that my work is embedded in the collaborative practices of an art/research/ecology collective called Gesturing Towards Decolonial Futures. Most of the ideas presented in this text emerged in the context of this collaborative work with other scholars, artists, students and

Indigenous knowledge keepers who are involved in this collective. The GTDF collective works at the interface of questions related to historical, systemic and ongoing violence and questions related to the unsustainability of our current habits of being that is based on exponential growth, overconsumption and the extraction of fossil fuels from a planet that is a finite living entity (Stein 2019; 2020). The pedagogical approach used in the collective's work is based on a non-Western form of psychoanalysis grounded in translations of insights and practices of Indigenous communities from Canada and Latin America. Part of the work of the GTDF collective is about mapping ethical pathways for forming relationships and collaborations with Indigenous communities based on trust, respect, consent and accountability, rather than extraction and appropriation (Jimmy, Andreotti, and Stein 2019; Whyte 2019). The work of the collective also involves artistic, embodied and land-based educational experiments.

Enduring challenges

The global health crisis of the COVID-19 pandemic has drawn attention to the fact that we are entering a period of intensification of multiple crises, including the cascading effects of inequalities, racial and colonial violence, climate crises and biodiversity loss, economic austerity, precarity and instability, mental health crises, political polarization, large-scale human migration, and more (Andreotti et al. 2018). While some still see the current pandemic as just a temporary interruption of a recoverable familiar normality, others, like Inuit artist Taqralik Partridge, caution that COVID-19 could be just the 'warning shots' of a major storm humanity will need to face together. From thisperspective, these juxtaposed crises will reshape 'normality' to an extent that is impossible to imagine at this point. We can predict that inequalities will likely be exacerbated and the ecological impact of these changes will also be significant. If we consider the waves of disruption and instability that were already unfolding before the global pandemic, and that were intensified by it, the phrase the 'end of the world as we know it' (Kingsnorth and Hine 2009; Ferreira Da Silva 2014) seems appropriate to describe our current circumstances. In this context of major traumatic disruptions in our cognitive, affective, relational, economic and ecological environments, how can education prepare all of us to engage in sober, sensible, creative and responsible ways with these challenges?

Education, within and beyond the parameters of formal schooling, has historically been tasked with steering learning towards objectives that could secure human survival as well as the reproduction of cultural norms and ideals. However, when the reproduction of cultural norms and ideals conflicts with the aim of securing human survival, we have a paradox and a problem. A few scholars associated with educational philosophy have started to engage with this issue. I briefly introduce four examples that are important for the context of this article.

Kopnina (2016, 2020) has problematized how the Sustainable Development Goals promote exponential economic growth as a highly problematic solution to inequalities and climate change. Wals (2020) has drawn attention to a hidden curriculum of unsustainability permeating mainstream education. The UNESCO background paper 'Learning to become with the world: Education for future survival' (Common Worlds Research Collective 2020) argues that education that separates humans from the rest of the world is both delusional and futile. Pradamos (2015) has critiqued the complicity of education in reproducing common place ideas about economic growth, technology and progress, arguing that education should focus on an alternative environmental ethics and on teaching alternatives to growth, such as the theory of degrowth. He states that 'in the current context of unacceptable inequality and ecological collapse we cannot continue teaching as if we were not losing from 40 to 200 species a day.' (Pradamos 2015, 153)

These examples illustrate that a more robust conversation about our discipline's complicity in the reproduction of unsustainability is necessary. Issues and questions raised in these articles that could guide this conversation include: What would be ethical to do if we accept that formal education is supporting an economic system that is undermining the conditions of possibility for the biophysical survival of human and other species? Is it ethical to assume and to foster hope in the continuity of the current system if this system is geared towards exceeding the capacity of the planet? Is it ethical to perpetuate the cultural norms and ideals that have brought us to the brink of extinction and possible social and environmental collapse? Can the cultural paradigm that created this situation also point us to a way out?

While the analyses presented so far are important, they also fail to substantially name and address the relationship between unsustainability and historical, systemic and on-going violence. This is not surprising, as it is rare to find discussions that address the relationship between climate change and multiple forms of colonial/imperial violence in the context of education. Scholars engaged with decolonial studies in Latin America attempt to explain why this erasure of violence is common. Mignolo (2000), for example, argues that modernity, as a project of a local (European) history with a global (imperial) design, is usually presented as a product of seamless human progress, merit, development and evolution, while hiding that violence and unsustainability are its conditions of possibility, or its underside. In order to emphasize the inseparability of modernity and colonial/imperial violence, the Latin American decolonial scholars use the term 'modernity/coloniality' to refer to modernity (Escobar 2007). Maldonado- Torres (2007) refers to coloniality as 'long-standing patterns of power that emerged as a result of colonialism, but that define culture, labor, intersubjective relations, and knowledge production well beyond the strict limits of colonial administrations' (2007, 243). I use the term modernity/coloniality as a pedagogical device to underscore the fact that violence and unsustainability are not collateral damages of modernity, but what make modernity possible.

When the production of violence that makes modernity/coloniality possible is invisibilized, well-meaning critiques and desires to reform existing systems often foreclose the fact that these existing systems have been created and are subsidized by historical, systemic and on-going harm (Andreotti 2011; Baker 2012). A foreclosure (Spivak 1999) is a form of socially sanctioned ignorance or denial – something we need to repress in order to justify our beliefs, hopes, desires and perceived entitlements.

Creating and imagining change from a space of socially sanctioned denials tend to reproduce harmful cognitive, affective and relational patterns that are rooted in the same old violent and unsustainable system.

The GTDF research/arts/ecology collective has mapped four denials that severely restrict the capacity of those of us socialized within modernity-coloniality to sense, relate and imagine otherwise (Stein et al. 2020b):

- the denial of systemic, historical and ongoing violence and of complicity in harm (the fact that our comforts, securities and enjoyments are subsidized by expropriation and exploitation somewhere else);
- the denial of the limits of the planet and of the unsustainability of modernity- coloniality (the fact that the finite earth-metabolism cannot sustain exponential growth, consumption, extraction, exploitation and expropriation indefinitely);
- the denial of entanglement (our insistence in seeing ourselves as separate from each other and the land, rather than 'entangled' within a living wider metabolism that is bio-intelligent); and,
- the denial of the magnitude and the complexity of the multiple problems we need to face together (the tendency to look for simplistic solutions that make us feel and look good and that may address symptoms, but not the root causes of our collective complex predicament)

Education for interrupting denials

The discipline of education within modern/colonial societies actively participates in the reproduction of these denials. Our discipline is not well equipped to interrupt denials since it mostly works with modern/colonial theories of Cartesian subjectivity focused on the mastery of content, and that virtually erase the role of our individual and collective unconscious. The standard mode of modern/colonial education is focused on addressing ignorance by 'filling up cups' with knowledge, competencies and skills. Another analogy that could be used for modern/colonial education is the image of someone climbing or conquering a peak, where students are prepared to arrive at a state of mastery, readiness and confidence for being functional in a given world. I will use the term 'mastery education' as a short hand to refer to the cluster of educational approaches focused on filling cups and conquering peaks.

Conversely, the type of psychoanalytic education that could address the denials presented before, which I provisionally call 'depth education', is very different because it requires a form of teaching that can surface problematic aspects of our individual and collective unconscious, which can enable us to work through what is unknowable, to relate beyond knowledge, identity and understanding (Andreotti, Biesta and Ahenakew 2011) and, to disarm our ego-logical defences (Biesta 2016). Depth education can develop our capacities, dispositions and stamina to hold space for difficult and painful engagements without feeling overwhelmed and immobilized or demanding to be rescued from discomfort. Depth education may also create the conditions where it is possible for us to accept Todd's (2015) invitation to face humanity in all its complexity, including the human wrongs that subsidize the dispensation of our 'human rights' (Spivak 2004).

This is difficult because 'mastery education' strengthens modern/colonial desires and attachments and creates educational expectations that render depth education counter-intuitive. In comparison to the images of cups and peaks that I evoked for positive education, 'depth education' is about transforming cups into onions and peeling their layers, or, rather than conquering peaks, one appropriate image would be deep diving into a freezing ocean without an oxygen tank in order to find out that we are both the water and the plastics and toxins that we (humans) dump into it.

In our field of education, there has also been an enduring debate about whether education should be about the transfer of content, which centers the teacher and a predefined curriculum, i.e. the 'banking' education that Paulo Freire (1970) has critiqued; or about the learner's construction of knowledge, where the teacher centers the learners' preferences and motivations, i.e. the learnification that Gert Biesta (2017) critiques. Both the centering of teachers and the centering of students privilege the codification of the world (by learners or teachers), rather than an encounter with an unknowable world on the world's terms (Biesta 2016). This distinction is important because what I am calling depth education, the kind of education that addresses disavowals and denials, centers neither the teacher, nor the learner, but the world itself. In order to center the world as a dynamic entity, or metabolism, the usual relationship with knowledge that turns the world into an object of knowledge production shifts to a relationship where everything and everyone are entangled as multi-layered moving entities who are extensions of the undefinable world, including knowledge itself (see Ferreira Da Silva 2016; Mika 2017; Stein et. al. 2020a).

The modern/colonial desire to know the world to control it (Gandhi 1999) prevents us from relating to the world in its full complexity, plurality, movement, entanglement and indetermination. For this desire to be interrupted, it first needs to be identified within ourselves and that is where images, metaphors, and strategic intellectual distinctions become indispensable in the context of depth education. For example, a distinction between the desire to 'learn from' something or someone and the disposition to 'be taught by' something, someone, or the world can be really useful to re-orient education towards (pedagogical) movement instead of form. As Biesta (2016), drawing on Levinas, has argued, in the desire to 'learn from', there is already an expectation and anticipation of what this learning will be about, while the disposition to 'be taught by' requires a renunciation of expectations and projections to make room for something unexpected that can interpolate us and move us in unpredictable ways.

The peeling of onion layers that happens in depth education also aims to address the limits of both Hegelian and Marxist dialectics, which can be brought together in the formula 'thesis versus anti-thesis = synthesis', which is characterized by the overcoming of contradictions and sustained by a teleology of progress. The coloniality of dialectics has been critiqued from different angles (Moore 1994; Mignolo and Dabashi 2015; Ciccariello-Maher 2017; Alpert 2019; Pratt and Yingnan 2019). Dussel (1994) has proposed the term 'analectics' as a substitute for dialectics (see also Kim 2019). With reference to depth education, I use the same term provisionally, with a different intent than Dussel (who is not focused on psychoanalytic approaches), to refer to the exploration of regimes of visibility, intelligibility, and affectability that delimit the boundaries of our representational possibilities within modernity/coloniality.

Analectics requires deconstructive motions, but its pedagogical intent goes beyond deconstruction, towards the noncoercive rearrangement of desires that Spivak (2004) talks about (see also Kapoor 2020), as well as what I call ontogenesis (Andreotti 2019). Ontogenesis is about shifting the ontological terms that enable/disable the folding/unfolding of existing and new possibilities.

I will attempt to illustrate how both terms, analectics and ontogenesis, would work in the field of education. Analectics, as I use it here, is at work in the critique of the romanticization of humanity in Sharon Todd's scholarship about cosmopolitanism (Todd 2015). Todd's critique examines the effects of the production of distinctions between what is 'human' and 'inhuman'. This distinction is at work when we talk about a 'dehumanizing' act, about someone 'losing humanity', or about the need to 'humanize' education, for example. When we make this distinction, we romanticize humanity by associating being human with virtue, beauty, goodness, greatness, benevolence and innocence while associating being inhuman with evil – or with whatever does not fit modernity's picture-perfect frame of humanity. There are numerous implications of this rather simple internalized distinction. Todd focuses on how it is at work when there is a call for education to 'cultivate' humanity, as Nussbaum (1998) suggests. The call to cultivate humanity is precisely a call to reproduce the romanticized idealization of humanity, grounded on the human versus inhuman problematic distinction. Todd proposes that, rather than cultivating humanity, education should interrupt the reproduction of this dichotomy and support people to 'face humanity' in all its spectrum. If we could, through education, indeed face the good, the bad, the ugly, the broken and the messed up of humanity (within and around us), a re-arrangement of desires would have taken place and ontogenesis (the clearing and opening of genuinely new onto-epistemic possibilities) could become a reality.

As a discipline, we tend to foreclose, for example, our discipline's relationship with colonialism, imperialism, and capitalism (Andreotti 2011; Stein 2019). It is interesting to note that, in the dialectic pattern we have socially internalized, the moment I suggest a critique of capitalism, I am already framed as proposing a predetermined alternative. Analectics focuses on the predetermination of this predetermination. If education was re-oriented towards ontogenesis, it could be opening up horizons of social-economic organization that go beyond what is imaginable within modernity/coloniality, beyond capitalism, socialism, communism, anarchism, green deals, doughnut economics, degrowth, or just transitions. Education tends to reproduce the architecture of imagining and being prescribed by modernity/coloniality and grounded on notions of human exceptionalism and the romanticization of humanity that Todd's work has examined. The way modernity/coloniality pre-figures our thought and imagination is faster than both thought and imagination themselves. These limits of the imagination, which is conscripted and constricted by modernity/coloniality, are explored further in the second part of this paper.

Education 2048 thought experiment

As an illustration of what I call depth education, the second part of this paper will present a thought experiment that was collaboratively designed by the GTDF collective.

Education 2048 is a thought experiment that was created for a conference in Germany in August 2020. The topic of the conference was utopian visions of the future and I was invited to present what I imagined schools to look like in 2048. The year was chosen to mark 100 years from the date that the Universal Declaration of Human Rights was adopted by the United Nation's general assembly.

There are two parts to this thought experiment. In the first part, the experiment invites you to imagine a scenario, in the second part, you are invited to consider two sets of questions. In the first part, the scenario places you on 10 December 2048. You are in a 3D virtual-reality world-skype call where people across the globe have gathered to decide the direction of education after a period of catastrophic events. These events have happened as a result of bad human choices followed by a period where humanity managed to be taught by the consequences of their actions and to co-exist in wiser and healthier ways. In the second part of the experiment, you are invited to consider questions about the scenario and questions related to your affective responses to the experiment itself. Since the conference in Germany, this thought experiment has been used a few times in educational conferences, therefore I will also make a brief comment about common responses from educational practitioners.

Depth education invites you to engage in deep imaginative diving that can prompt cognitive and affective dissonance, and, to observe the patterns of your own responses to this invitation. Unlike mastery education approaches, this depth education experiment is not trying to impart content or to offer a specific teleology – a point of arrival that we all should agree with. The exercise is neither descriptive or prescriptive, instead it intends to clear space and build capacity for us to reflect on the role of education in confronting the potential and/or likelihood of social and ecological collapse in (and/or beyond) our lifetimes. The exercise invites you to 'zoom out', taking a step away from yourself, and to observe yourself observing, with curiosity, without investing emotionally in the reactions or resistances that may emerge towards the exercise.

Education 2048

Imagine today is 10 December 2048 and you are participating in a 3D virtual reality world-skype call with the intent to decide how we should educate our children and ourselves after the devastating impact of the events of the past 30 years, which were caused by human greed, arrogance, stubbornness and unwillingness to learn from human wrongs. For our consultation, it is important that we remember: a) what education is for; b) what the past has taught us; c) what made a difference to get us out of the mess; d) what we know now; and e) what we are still being taught. We will review these together before starting our consultation. There are one billion people on this call, a third of the world's total population today. We will start by recalling significant events that happened in the last three decades. What you will see next is the text of this presentation only, the associated images and videos are now being transferred to your body implants, please blink twice to accept the files.

The period between 2018 and 2027

The period between 2018 and 2027 was marked by an interruption of our dreams of prosperity-as-consumerism. This period started with school climate strikes, young people seemed to be aware that the path we had chosen was both violent and unsustainable, they already knew that their generation would not have the same opportunities for social mobility that their parents and grandparents enjoyed, nor the stability and well-being afforded by their class in relatively healthy ecosystems. In 2020, the Covid-19 global pandemic hit, followed by several outbreaks of different strains of the virus. Racial tensions and civil unrest also marked this period, especially in highly divided and volatile countries like Brazil, Hong Kong, UK and the USA.

From 2022, we had a mounting economic crisis, mass unemployment, mass homelessness, mass migration, localized famines, and the intensification of social and racial inequalities. Reactionary populist governments were democratically elected all over the world by harnessing the power of hyper-political polarization and fuelling antagonism and resentment. Extreme weather, flooding and catastrophic forest fires became routine. Pollinator populations collapsed globally in 2026. In 2019, we thought there was a silver lining to all this as we celebrated our collective efforts when we registered the smallest size of the hole in the ozone layer on record, but in 2020, the hole appeared deeper and larger than it was first discovered. For a few months of that same year, carbon emissions noticeably dropped in places where coronavirus lockdowns were observed. At the beginning of the pandemic there was much excitement that people's attitudes towards fossil fuel travel had changed forever, but very soon emissions soared again. By 2027, alpine skiing was only possible in the Himalayas.

In this 10-year period, schools became even more obvious sites of political and economic struggle. Governments in many countries, including the USA and the UK, forbade teachers to talk about racial privilege or the complexities of gender and sexual identities. Teachers were also prohibited to present anti-capitalist perspectives or perspectives that challenged accounts of history authorized by the state. However, despite political disputes, the protection of unrealistic economic interests was still prioritized in the curriculum: at large, schools were still teaching a learner centered, competencies based, technology and market-oriented curriculum driven by hyper- individualistic metropolitan consumerist values and principles.

By this time, curriculum, examination and school social services had largely been outsourced to private companies. In 2026, a well-known corporation in the US that had been lobbying the government for decades offered a more cost-effective option for online global education and teacher training. They standardized a global curriculum powered by artificial intelligence, causing mass teacher unemployment. Governments saved trillions in tax dollars but could not re-invest in social services because the funds were reallocated to offset the historical level of debt created by the coronavirus stimulus packages. This shift in capital created the single greatest increase in wealth inequality in historical records, with the top 0.01% amassing the equivalent net worth of 90% of the world's population.

The period between 2028 and 2037

The period between 2028 and 2037 was marked by the real threat of human extinction. By 2028, the Amazon forest was down to a third of its size; the remainder teetered on the edge of becoming a biologically inert desert, similar to the exhausted agricultural fields of the US corn-belt. Ironically, the market share of Jeff Bezos' Amazon company was worth a third of the total global wealth. The average temperature rose 2.47 degrees Celsius across the globe, previous mitigation efforts had been too little too late. All permanent ice in the Arctic was lost. Children read about polar bears and many other arctic species in the new histories of extinction; sad companions to the images and stories of mammoths, sabre-tooth tigers, and giant land sloths that filled children's books in the late 20th Century.

All permafrost in Siberia melted, releasing methane and unleashing viruses and fungal strains that we were not immune to or prepared for. We had extensive desertification that led to massive crop failures that, in turn, caused the global food supply chain to break down. This resulted in the worst global famine ever experienced. There were also disastrous typhoons, hurricanes and tsunamis making large areas uninhabitable. Many people were displaced. Borders were closed and militarized because of mass migration. Most states failed to sustain their welfare safety net and many people were left without any government support for healthcare, sanitation, or waste management. Pensions and unemployment benefits also became a thing of the past as systems of wage-labor foundered, and mid-human-range, future forecasting had become impossible. Reactionary dictatorships were installed. Governments used mass surveillance and AI technology, wrapped in the rhetoric of 'maintaining law and order', to sweep away civil rights and freedoms and control citizen mobility. Torture and genocide against people and groups perceived as inconvenient increased at an alarming rate and became a normalized extension of police and military tactics. Nuclear and biological-weapons were unleashed in wars.

Five factors contributed to unparalleled loss of human life during this time: 1) unprecedented famines; 2) major viral and fungal outbreaks; 3) a global mental health crisis; 4) incurable new diseases caused by combined toxins and microplastics in food and water; 5) violent civil conflicts, including state sanctioned violence, terrorism and police brutality. Needless to say, Black, Indigenous and racialized populations were disproportionately targeted by state-sponsored violence and suffered the most in these events due to unparalleled levels of inequality. It is important to acknowledge that many of these people and communities had been experiencing these levels of trauma for generations, and had been sounding the alarm for ages, but most of us were indifferent and focused on trying to sustain our way of life in spite of all evidence that it was deadly.

Online schooling became unviable in 2033, and the corporation that had taken over global online learning went bankrupt in 2034, although the C-suite executives received very generous pay-outs, bonuses, and severance packages prior to dissolution. The global economy completely collapsed in 2036, a year after Elon Musk started a human colony on Mars.

The period between 2038 and 2047

In the period between 2038 and 2047, we finally accepted that we were part of the problem and that we needed to get our act together, grow up and engage with our messy and painful reality to avoid being wiped out. The Mars colony tragically failed in 2038 destroying our hopes for life on another planet. In 2039, a massive event made us all suddenly realize that we had messed up big time. We realized that we were addicted to arrogance, consumption, unaccountable autonomy and control. We accepted that we needed mass rehabilitation. We grasped the gravity of the fact that we were only three billion people left on the planet. We understood that we had caused the extinction of 70% of all species – and the extinction of all life in whole parts of the Earth – and we were extremely close to causing our own. We recognized that planet Earth is alive and we are part of its metabolism, not the center of the world, or a special species. We also worked out that humanity is capable of both horrendous and wonderful things. We started to face our own and others' humanity in all its complexity and to be taught by the human wrongs we had inflicted upon each other, upon other beings and upon the planet. Then we all had to learn quickly, collectively and without schools or moral manifestos:

- to heal intellectually, emotionally, relationally, economically, ecologically and politically;
- to abolish colonial and racial violence, inequality, hierarchies of worth and separations;
- to center the Earth and de-center our egos, identities, human narratives and projections;
- to age and to die in generative ways;
- to care for everything and everyone, rather than compete;
- to plant, to repurpose technology, to compost, to repair and to regenerate everything;
- to prioritize the common good for humans, non-humans and the planet;
- to use words and conversation carefully and wisely, with humility and maturity;
- to own up, to sober up, to clean up, to grow up, to show up and to exist differently.

10 December 2048

So today, 10 December 2048, we convene to decide how to educate our children for human responsibility considering the needs of the next seven generations of humans and non-humans alike. We need our children to learn from human wrongs: from the violence and unsustainability caused by humanity, from our repeated mistakes of the past so that they can only make different mistakes in their future. From the day children are born, their education should prepare them to become healthy and wise elders and good ancestors for all relations. We cannot afford to repeat history. Today we decide how to do this together, as a planet-wide human and non-human family. This is the end of the scenario.

Questions

In the second part of thought experiment, participants are invited to engage with sets of questions organized in two groups: 'dipping-in' questions that engage with the scenario itself, and 'deeper-dive' questions that engage with the cognitive and affective frames that informed how people might have engaged with the scenario. A sample of these questions is presented below.

Dipping-in questions include:

(1) What do you think was the event that made us realize we needed another way to (co)exist?
(2) What motivated us to keep going on an unsustainable and violent path until 2039? How were policies, technologies and market economies used to support violence and unsustainability?
(3) What motivated us to keep going on a different path after 2039? What healthier and wiser forms of organization and economies emerged? How was technology used to support a different form of (co)existence?
(4) Who were the people who survived? Who perished? Why?
(5) What did we fail to learn up to 2018 that could have prevented the most harmful events of the next two decades?
(6) Knowing what you know in 2048, if you could go back to 2020, what advice would you give to the people who are about to face the events that unfolded?

Diving-deeper questions include:

(1) What affective responses emerged during this exercise? Anxiety? Sadness? Resistance? Relief? Exhaustion? Excitement? Hope? Hopelessness? Anger? Frustration? Defensiveness? Loss? Grief?
(2) The collapse described in the scenario has already happened for many people and other species. In many cases, the 'collapse' they face on a daily basis is what maintains the comforts, enjoyments, securities and conveniences of those of us who have so far been protected from it. How do you relate to your own complicity in violence and unsustainability? What sensations emerge in your body when you reflect on this?
(3) How did you respond to the portrayal of humanity as arrogant, stubborn and unwilling to learn? Who would likely see humanity from this perspective? Did you witness a desire for humanity to be portrayed only as beautiful, brave, wise, loving and kind?

(4) Did you witness a desire for hope in the continuity of our current system and reassurance that everything would be okay?
(5) What realizations (if any) have you had with this exercise?

As members of my research collective presented this experiment in different educational contexts, we observed and mapped four interesting clusters of responses that we found were relevant to further our inquiry into depth education. The first cluster of responses was characterized by relief that the experiment addressed frustrations related to the fact that these topics are rarely addressed in education. Responses in this cluster expressed relief (with the articulation of the catastrophic unfolding of events), anger (at our collective inability to see this already happening), excitement (that we are finally talking about it), and scepticism (in relation to the awakening presented as a conclusion).

The second cluster was characterized by initial frustration followed by hope: frustration with the description of events, but hope with its conclusion. Responses in this cluster reflected surprise (with the exercise), anxiety (with the unfolding catastrophe), hope (with the conclusion), and concern (for the length of time it could take for solutions to be found).

The third cluster was characterized by immobilization: responses indicated that the experiment had pushed people beyond their 'stretch' zones into zones of panic or immobilization. Responses in this cluster reflected resistance (to the exercise), opposition (to the existence of the exercise), fear (of hopelessness caused by the exercise), despair (with the scenario), feeling overwhelmed (with the complexity of the situation), and denial (of complicity in the problems presented).

The fourth cluster was characterized by signs of re-arrangement of desires. Responses in this cluster reflected bewilderment (at the scale of the problem), disorientation and dis-identification with one's own priorities (loss of satisfaction with familiar cognitive/affective frameworks), un-ease with leaving this open coupled with motivation to keep working through complexities, feeling ill-equipped to take next steps (but not feeling immobilized), and feelings described as 'open heart surgery' and 'led ball in the stomach'.

It is important to acknowledge that images of apocalyptic futures have been used by people across the political spectrum, including the far-right, to mobilize different kinds of collective action. It is also useful to acknowledge that within modernity/coloniality we are expected to trust a positive image of the future, therefore, people who believe in this premise may argue that bringing up the potential of a catastrophic future leads to alienation, immobilization, indulgent and irresponsible behaviour, and/or the reproduction of a non-relational ethics (see Fagam 2017, 2019). Depth education is not about mobilizing a specific action, but about addressing denials and noncoercively rearranging socially sanctioned modern/colonial desires. It is also about clearing the space and expanding capacities and dispositions for us to imagine, sense, and relate otherwise, focusing on pedagogical movement, rather than a pedagogical form or point of arrival. The Indigenous communities the GTDF collective collaborates with have extensive experience in creating land-based education oriented towards these goals. With our imperfect and agonistic translations of this pedagogical orientation into Western contexts, through pedagogical experiments like Education 2048, we are just beginning to understand how this could work in practice.

To conclude, I would like to invite us all as philosophers of education and educational practitioners from the global north and south to consider our roles and responsibilities as we prepare ourselves and others to confront a multitude of crises. Could education as we know it equip humanity to hold space for difficult (painful, overwhelming, irritating) issues and conversations, away from desires to be rescued/coddled and without demanding quick fixes? Could education as we know it be re-oriented away from metropolitan consumerist individualism? Could education as we know it prepare us to show up with maturity, sobriety, humility and accountability as we face storms together, without hurting (or killing) each other or further harming (or fully destroying) the planet and other non-human beings? Could education as we know it equip us to face the potential or likelihood of the end of the world as we know it, or do we need something else?

Acknowledgements

I would also like to thank the Musagetes Foundations for the generous support offered to the GTDF collective.

Disclosure statement

No potential conflict of interest was reported by the author.

Funding

This article is written in the context of three research projects funded by the Social Sciences and Humanities Council of Canada and Mitacs: "Social Innovations in Education for Decolonial Futures", "Braiding new possibilities for co-existence: holding spaces for difficult conversations", and "Re-imagining Aboriginal education for a shared future".

ORCID

Vanessa De Oliveira Andreotti http://orcid.org/0000-0003-1608-2488

References

Alpert, A. 2019. *Global Origins of the Modern Self, from Montaigne to Suzuki*. Albany: SUNY Press.

Andreotti, V. 2011. *Actionable Postcolonial Theory in Education*. New York: Palgrave Macmillan.

Andreotti, V. 2019. "The Enduring Challenges of Collective Onto-(and Neuro-) Genesis." *Lapiz (Latin American Philosophy of Education Society Journal)* 4: 61–78.

Andreotti, V., G. Biesta, and C. Ahenakew. 2011. "Between the Nation and the Globe: Education for Global Mindedness in Finland." *Globalisation, Societies and Education* 13 (2): 246–259.

Andreotti, V., S. Stein, A. Sutherland, K. Pashby, R. Suša, and S. Amsler. 2018. "Mobilising Different Conversations about Global Justice in Education: Toward Alternative Futures in Uncertain Times." *Policy & Practice: A Development Education Review* 26 (Spring): 9–41.

Baker, M. 2012. "Modernity/Coloniality and Eurocentric Education: Towards a post-Occidental Self-understanding of the Present." *Policy Futures in Education* 10 (1): 4–22.

Biesta, G. 2016. "The Rediscovery of Teaching: On Robot Vacuum Cleaners, Non- Egological Education and the Limits of the Hermeneutical World View." *Educational Philosophy and Theory* 48 (4): 374–392.

Biesta, G. 2017. *The Rediscovery of Teaching*. Abingdon: Taylor & Francis.

Ciccariello-Maher, G. 2017. *Decolonizing Dialectics*. Durham: Duke University Press.

Common Worlds Research Collective. 2020. "Learning to Become with the World: Education for Future Survival, UNESCO Background Paper for the Futures of Education Initiative." Accessed 31 October 2020. https://unesdoc.unesco.org/ark:/48223/pf0000374032

Da Silva, F. 2014. "Toward a Black Feminist Poethics: The Quest(ion) of Blackness toward the End of the World." *The Black Scholar* 44 (2): 81–97.

Da Silva, F., 2016. "On Difference without Separability." *32nd Bienal De São Paulo/Art Biennial, Incerteza viva exhibition catalogue* (57-65). São Paulo: Fundação Bienal de São Paulo.

Dussel, E. 1994. "Leopoldo Zea's Project of a Philosophy of Latin American History." In *Latin American Identity and Constructions of Difference*, edited by A. Chanady. Translated by Amaryll Chanady, 26–42. Minneapolis: University of Minnesota Press.

Escobar, A. 2007. "Worlds and Knowledges Otherwise: The Latin American Modernity/coloniality Research Program." *Cultural Studies* 21 (2–3): 179–210.

Fagan, M. 2017. "Who's Afraid of the Ecological Apocalypse? Climate Change and the Production of the Ethical Subject." *The British Journal of Politics and International Relations* 19 (2): 225–244.

Fagan, M. 2019. "On the Dangers of an Anthropocene Epoch: Geological Time, Political Time and Post-human Politics." *Political Geography* 70: 55–63.

Freire, P. 1970. *Pedagogy of the Oppressed*. London: Continuum.

Gandhi, L. 1999. *Postcolonial Theory: A Critical Introduction*. New Delhi: Oxford University Press.

Jimmy, E., V. Andreotti, and S. Stein. 2019. *Towards Braiding*. Guelph: Musagetes Foundation.

Kapoor, I. 2020. *Confronting Desire: Psychoanalysis and International Development*. Ithaca: Cornell University Press.

Kim, D. H. 2019. "Alterity, Analectics, and the Challenges of Epistemic Decolonization." *The Southern Journal of Philosophy* 57: 37–62.

Kingsnorth, P., and D. Hine. 2009. "The Dark Mountain Manifesto." Accessed 31 October 2020. http://dark-mountain.net/about/manifesto

Kopnina, H. 2016. "The Victims of Unsustainability: A Challenge to Sustainable Development Goals." *International Journal of Sustainable Development & World Ecology* 23 (2): 113–121.

Kopnina, H. 2020. "Education for the Future? Critical Evaluation of Education for Sustainable Development Goals." *The Journal of Environmental Education* 51 (4): 280–291.

Maldonado-Torres, N. 2007. "On the Coloniality of Being: Contributions to the Development of a Concept." *Cultural Studies* 21 (2–3): 240–270.

Mignolo, W. 2000. *Local Histories/global Designs: Coloniality, Subaltern Knowledges, and Border Thinking*. New Jersey: Princeton University Press.

Mignolo, W., and H. Dabashi. 2015. *Can Non-Europeans Think?* London: Zed Books.

Mika, C. 2017. *Indigenous Education and the Metaphysics of Presence: A Worlded Philosophy*. Abington: Taylor & Francis.

Moore, D. L. 1994. "Decolonializing Criticism: Reading Dialectics and Dialogics in Native American Literatures." *Studies in American Indian Literatures* 6 (4): 7–35.

Nussbaum, M. C. 1998. *Cultivating Humanity*. Cambridge, MA: Harvard University Press.

Prádanos, L. I. 2015. "The Pedagogy of Degrowth: Teaching Hispanic Studies in the Age of Social Inequality and Ecological Collapse." *Arizona Journal of Hispanic Cultural Studies* 19: 153–168.

Pratt, J., and Y. Zhao. 2019. "A Daoist Critique of Dialectics and Why It Matters." Accessed 31 October 2020. doi: 10.2139/ssrn.3140946.

Spivak, G. C. 1999. *A Critique of Postcolonial Reason: Toward A History of the Vanishing Present*. Cambridge, MA: Harvard University Press.

Spivak, G. C. 2004. "Righting Wrongs." *The South Atlantic Quarterly* 103 (2): 523–581.

Stein, S. 2019. "The Ethical and Ecological Limits of Sustainability: A Decolonial Approach to Climate Change in Higher Education." *Australian Journal of Environmental Education* 35 (3): 198–212.

Stein, S., V. Andreotti, R. Suša, C. Ahenakew, and Č. Tereza. 2020a. "From "Education for Sustainable Development" to "Education for the End of the World as We Know It"." *Educational Philosophy and Theory* 1–14. doi:10.1080/00131857.2020.1835646.

Stein, S., V. Andreotti, R. Suša, S. Amsler, D. Hunt, C. Ahenakew, E. Jimmy, et al. 2020b. "Gesturing Towards Decolonial Futures: Lessons Learned Thus Far." *Nordic Journal of Comparative and International Education (NJCIE)* 4 (1): 43–65.

Todd, S. 2015. *Toward an Imperfect Education: Facing Humanity, Rethinking Cosmopolitanism*. Abington: Routledge.

Wals, A. E. J. 2020. "Transgressing the Hidden Curriculum of Unsustainability: Towards a Relational Pedagogy of Hope." *Educational Philosophy and Theory* 52 (8): 825–826. doi:10.1080/00131857.2019.1676490.

Whyte, K. 2019. "Too Late for Indigenous Climate Justice: Ecological and Relational Tipping Points." *Wiley Interdisciplinary Reviews: Climate Change* 11: 1. doi:10.1002/wcc.603.

OPEN ACCESS

'Landing on Earth': An educational project for the present—A response to Vanessa Andreotti

Sharon Todd

ABSTRACT
This paper responds to Vanessa Andreotti's keynote address. In it, I draw out some educational implications of facing the everyday denials of the climate emergency. In particular, I mobilise Bruno Latour's phrase 'landing on Earth' to indicate that the very terms through which we understand education, particularly as it relates to the future, require a profound shift.

> As everyone learned in school, when our idea of the position of the Earth in the cosmos is modified, a revolution in the social order may ensue. Remember Galileo: when astronomers declared that the Earth moves around the Sun, it felt as though the whole fabric of society was under attack. Latour and Weibel (2020, 12)

Vanessa Andreotti speaks with prescient clarity about the afflictions borne by all forms of planetary life at this very moment. Whether writing about the end of the Anthropocene or calling for a new politics of the Critical Zone, scholars from across all disciplines have underscored what Andreotti here has called the 'end of the world as we know it'. What is so striking is that it is a refrain that is not only of this time of climate crisis, but comes from a very old song. 'The end of the world as we know it' is not new for indigenous peoples, for species already extinct, for languages and cultures lost, for those who have been enslaved or whose homes and lives have perished through the effects of environmental neglect and colonisation (which have gone hand in hand), or for the disappearance in some parts of the world of the very plants, trees, ice, clean air and water that make life possible on this Earth.

The denials that Andreotti speaks of are further evidence of how difficult it is to hear, to really listen to, the roar of this loss. And it is not only those committed to authoritarian post-truth politics that have difficulty facing reality and retreat into denial. Each one of us, at least in the global north, no matter our political affiliations or habitus, constantly dances at the edge of these denials, at times out of wishful thinking, grief, paralysis or fear. We dance at their edge, because words fail us, and they fail us, because we cannot conjure a future with any surety. Our familiar habits of observation, description, and explanation are inadequate for dealing with the enormity of not only factual information, but the affective displacement that accompanies our exposure to it. Even Tim Flannery (2020), an eloquent, impassioned and award-winning author and environmentalist, is at a loss for words when it comes to talking about the climate crisis to his seven-year-old son; not because his son cannot understand the meaning of the devastation, but because he understands all too well. What is at stake in these intimate exchanges is nothing less than the next generation's ability to survive, his own son's ability to survive. In the face of this, we can see how denial comes easier than words.

However, these forms of denial that Andreotti speaks of undermine our capacity for repair and restitution – dare I say atonement – since they are so strongly internalised as part of our inner landscape, with all its flaws as well as its rational powers. The denial, for instance, to see ourselves as entangled (Andreotti's third denial), inextricably and persistently, with other life forms, is founded on a deep cut – or wound – that separates what we call 'me' from the necessary multiplicity of 'others' with whom life is made possible – from the bacteria within our bodies, to the organisms that constitute much of the food we eat, from the minerals and salt we depend on from the earth and sea, to the composition of the air we breathe (to say nothing of the viruses that need us to survive). Understood in this sense, our denial of entanglement is not merely a conceptual refutation but is an act of necrophilia – which is not about facing death which would be to see death as part of life, but more about a drive toward death as an outright denial of the living world – a denial that is quickly pushing us to the brink of extinction.

When I say that 'we' dance at the edge of denial, this is to suggest the subtle ways denial can work upon all of us; our political commitments, our philosophical outlooks, our worldviews are consistently challenged by the way we live our lives in the global north. We live lives of compromise, both in terms of the decisions we make and those that are made for us: the CO2 emissions for flights we take to see family; the decision to buy, use, and recycle plastics and seeing how plastics are destroying our oceans; our food choices that rely on unsustainable agricultural practices; the amount and kinds of energy we use and the poisoning of the air we breathe; the extent to which our toxic waste from our phones and computers ends up on doorsteps of communities we don't ever have to face. These are the uncomfortable and disarming complexities to which Andreotti draws our attention. One's complicity is thus difficult to bear because it entails violence on a grand scale, reinforcing centuries of colonisation of people, of animals, of land that modernity's so-called progress has depended upon. No wonder we dance at denial's edge!

There now exist a proliferation of terms for living in and with this age of loss and complicity: eco-anxiety, climate depression, climate sorrow, climate grief, and solastalgia, to name a few. Coming to face these states of being is a project

Andreotti identifies as analectic and ontogenetic, a project that disrupts our desire to override sitting with complexities and paradoxes. While Gayatri Spivak (2004) has argued that education is about the 'non-coercive rearrangement of desire,' Andreotti extends this further in suggesting that we need to face our current desires and investments in order to make any real change or rearrangement possible. While this may be non-coercive work in the sense of non-violent, it is nonetheless difficult and painful. For Andreotti, this work constitutes what she calls a 'depth education' or a 'non-western psychoanalysis' – an education that analyses the conditions under which we continue to live life as if nothing is happening and turn it into an opportunity for profound realignment of our living arrangements, politics, and ethical commitments. For Andreotti, education cannot afford to be based on presuming the continuity of the existing system, with all our comforts, acts of disposal, and unequal distribution of who gets to live and how, but one that opens up possibilities for students, at all levels, to imagine and create alternatives. To think themselves and the world anew.

Simply educating for a system that is so out of joint with the consequences of environmental violence indeed is educating for denial – and irrelevance.

So what is the option? Andreotti introduces us to a thought experiment that brings us into 2048, into a future that now no longer seems entirely fictional, but *unheimlich*. There is thus a familiarity to the uncanniness experienced when we look into the gaping maw that is the future – a space where life is beyond precarity and a time that is seeking its own renewal.

When asked what kind of education system is now desirable, Andreotti is neither posing a question of artifice nor one that is easy to answer, for in order to respond to the question, there is a need to reflect on our own conditioning and capacities for moving beyond the given. It requires a shift, not unlike Galileo's, a radical decentring of ourselves and a radical centring of the world.

But of what does this world consist? As Latour (2017) puts it, the west has been built on seeing the world in two ways: the world we live *in* and the world we live *from*. The world we live in – the social, scientific, intellectual, and material ways we go about our everyday – is now colliding with the world we live from – the air, water, plants, animals, bacteria and other micro-organisms upon which we depend. The two worlds, traditionally separated, like the wound mentioned above, can no longer be held apart. For Latour (2017), this is a truly revolutionary moment, a moment to face Gaia; not to face the world as one organism, but the world as a network of pulsating forms of Life. Gaia, as Latour drawing on James Lovelock conceives it, is not a unity but a proliferation of Life itself, including the co-relation between animate forms and 'many other participants not usually counted in its balance sheet – atmosphere, soil, rocks, seas clouds, minerals, continents – that have been transformed, mobilized, generated, inhabited, engineered by life forms over eons of time' (Latour and Weibel 2020, 18). It is Life as it occurs in the Critical Zone – that is, those few kilometres that make up the surface of the Earth: from the depth of bodies of water to the atmosphere above the mountain peaks. Life forms alter materia, materia alters life. At one level we 'know' this from archeology, geology and oceanography, and through wisdom traditions and cosmologies, as well as through the effects fertilizer has on our house plants – it is a deep entanglement that defies our strategies for denial. Yet, on another level, we have not yet made this Critical Zone our home.

Latour, in a slightly different but complementary vein from Andreotti, thereby calls on us to 'land on Earth', to land in the Critical Zone, as a singular instance of Gaia that draws together the two worlds that have been separated. We, at least in the west and global north, have not lived on Earth recently; we are like strangers to our own home – another experience of *unheimlich*, but this time not one from 2048, but from 2020. The present is now the future, or is the future now the present? Moreover, we no longer have the luxury not to face it. "This historical moment – rendered earlier by euphemisms such as 'ecological crisis' or 'climate change' – now would best be taken as an existential crisis, a matter of life and death" (Latour and Weibel 2020, 13). How do we create an education for 'landing on Earth', where the world we live in and the world we live from come together?

As Andreotti observes, this requires activating the imagination, along with creating new sensibilities, new frameworks through which we can embody life differently in order to decolonise and generate alternative conditions of living with others, and recognise the deep interdependency that is Life itself. And, it also requires developing new languages, so that Earth becomes a place of home, a place of significance where the rent of separation can be mended. As Wall Kimmerer (2013) writes, 'to be native to a place we must learn to speak its language' (48). We need to speak anew, to listen to the language of Earth, not as some romantic recuperation of a lost past, or as a way of moving forward in a straight line of progression, but as a way of deepening a sense of place. We are at a loss for words because we are out of place.

But if landing on Earth is to be more than a metaphor, it needs to reinvent the very terms of education, not as a preparation for the world as it is, nor for some pre-defined future, but as a foray into what it could be. I see this as a profoundly aesthetic act, a way of engaging sensibilities. Aesthetics, as Latour and Weibel (2020) put it, 'renders one sensitive to the existence of other ways of life . . . and the shape of things to come' (19). Whether through conventional art practices, acts of imagination and creativity, thought experiments and the generation of new sensibilities, education is nothing less than that which supports a 'revolution of the social order' to return to the epigraph, something we have all been taught in school, but have forgotten. This, to me, is what Andreotti's depth education is calling for: a re-membering of future possibilities by facing complexities in the present.

Disclosure statement

No potential conflict of interest was reported by the author.

References

Flannery, T. 2020. "We Need to Talk to Our Kids about the Climate Crisis. But Courage Fails Me When I Look at My Son." *The Guardian*, November 3.

Latour, B. 2017. *Facing Gaia: Eight Lectures on the New Climatic Regime*. Cambridge: Polity Press.

Latour, B., and P. Weibel. 2020. "Seven Objections against Landing on Earth." In *Critical Zones: The Science and Politics of Landing on Earth*, edited by B. Latour and P. Weibel, 12–19. Boston: MIT Press.

Spivak, G. 2004. "Righting Wrongs." In *The South Atlantic Quarterly*, no. 103. 523–581.

Wall Kimmerer, R. 2013. *Braiding Sweetgrass: Indigenous Wisdom, Scientific Knowledge and the Teachings of Plants*. Kindle e-book. London: Penguin.

OPEN ACCESS

Education, anthropocentrism, and interspecies sustainability: Confronting institutional anxieties in omnicidal times

Helena Pedersen

ABSTRACT

Deborah Britzman's remarkable question, 'What holds education back?', appears more urgent than ever in a world of accelerating environmental crises, climate change, and what has been described as *omnicide* – the annihilation of everything. What, then, holds education back from initiating radical change under these urgent conditions? This paper introduces the notion of 'institutional anxiety' as a consolidating force and explores how it may condition possibilities for resistance. Bringing examples from ethnographic fieldwork and experiences of course development in conversation with psychoanalytic and schizoanalytic thought, a key catalyst of institutional anxiety is discussed: Anxiety related to 'the question of the animal' as a threat to human exceptionalism in educational practice and research. Confronting these anxieties could open new modes of being and acting in academic space and give interspecies ethics, justice and sustainability a chance to develop in omnicidal times.

Introduction: omnicide and the problem of education

In their recent editorial for the journal *Environmental Education Research*, Constance Russell and Reingard Spannring (2019) note that '[i]t has become a cliché to begin an editorial for this journal with a rehearsal of the challenges facing the planet and all its inhabitants' (1137). This keynote wrestles with a similar dilemma: How is it possible to invite a debate on the conference theme 'Education, the Environment and Sustainability' without resorting to repetitions, rehearsals, and clichés on the precarious state of the world? This is not a simple task, and in this paper, I don't claim to avoid clichés; I will rather underscore the significance of their messages. But I will also discuss education's position in the climate and environmental crisis by looking at the internal disposition of education, asking what holds education back (Britzman 2003) from becoming a transformative force for change. To this end, I will introduce the concept of institutional anxiety as *one* possible barrier against radical educational change. I will then explore how unspoken assumptions about human exceptionalism[1] (Haraway 2008), underpinning the anthropocentric infrastructure of educational institutions, become an anxiety-producing force whenever this exceptionalism is challenged, even in subtle and minor ways. I will further suggest that the momentary breakdowns in education caused by minor gestures of challenge and resistance (Manning 2016) hold a potential for being and acting differently in spaces of teaching and learning. In this discussion, I will draw on examples from my own empirical fieldwork.

But let me begin in another place: Danielle Celermajer's (2020) opinion piece 'Omnicide: Who is responsible for the gravest of all crimes?', written in the aftermath of the devastating Australian bushfires spring 2020, when over a billion animals (invertebrates excluded) are estimated to have been killed (Wood 2020). Escalating climate crisis caused by global warming, species extinction at alarming rates, ocean pollution, threats of ecosystem collapse and pandemic outbreaks originating in destructive human-animal relations, are crises that cannot fully be understood in terms of genocide or ecocide. Rather, *omnicide* – the annihilation of everything – may be the appropriate term for what is going on (Celermajer 2020). None of us, writes Celermajer, 'developed a specific intent to kill everything. But all of us have created and are creating the conditions in which omnicide is inevitable.' Who then, among all the human 'we' on this planet, have contributed most to these omnicidal conditions, is a question that has been discussed for years in the Anthropocene vs. Capitalocene debate (Chakrabarty 2009; Malm and Hornborg 2014), but among all the arguments that I may be guilty of rehearsing in this keynote, this is not one of them.

Instead, I want to turn to education. What is the responsibility of, and what can education achieve in omnicidal times? If I were a visiting alien trying to learn something from the complex institutional infrastructures on Earth, I would contend that education seems to be uniquely placed to contribute to the work of preventing terrestrial omnicide and direct the world's human and nonhuman inhabitants toward a less gloomy future. Research-based suggestions of how to begin this educational work are already available. A key issue in this work is how we relate to nonhuman animals. The open access 'Sydney Declaration on Interspecies Sustainability' (Probyn-Rapsey et al. 2016) argues for an expanded definition of sustainability to include interspecies ethics acknowledging that also animals have a right to the social, material and ecological bases for flourishing lives, sustained over time. The Sydney Declaration critiques anthropocentric notions of sustainability that

> ... stigmatize[s] animals as yet one more problematic source of GHG emissions, like cars or cement factories, rather than recognizing them as subjects who, like us, must cope

with the devastating impacts of climate change. Non-human animals aren't the cause of GHG emissions; *human exploitation of them is*. (Probyn-Rapsey et al. 2016, 137; emphasis in original)

The Sydney Declaration presents the following five calls to action to guide university policies for sustainable campuses:

(1) A university defines carbon footprint to include the contribution made by animal agriculture
(2) A university defines environmental sustainability to include considerations of food practices
(3) A university commits to producing knowledge to educate students and the broader community on the key contributors to climate change and areas for action (energy, transport, animal agriculture)
(4) A university ensures that its social justice commitments include interspecies ethics
(5) A university promotes plant-based diets (understood here as food free of animal products) to optimize staff and student health and wellbeing and supports individual choice and transition via institutional structures, campaigns and incentives (Probyn-Rapsey et al. 2016, 138).

Despite a few notable examples of UK and US-based universities that have taken policy initiatives in this direction (Probyn-Rapsey et al. 2016), institutional resistance to such initiatives remains profound. As research in critical animal studies has elucidated, the infrastructure of (at least Western) formal education rests on anthropocentric, or, to borrow Donna Haraway's (2008) term, human exceptionalist grounds, designed to privilege human interests while marginalizing or exploiting other animals and species through, for instance, science based on knowledge produced by animal use (e.g., Best 2009; Cole and Stewart 2014; Corman and Vandrovcová 2014; Gunnarsson Dinker and Pedersen 2016; Kahn 2003; Kahn and Humes 2009; MacCormack 2013; Nocella II et al. 2019; Oakley 2019; Wallin 2014). Coupled with the embeddedness of education in the animal-industrial complex (Pedersen 2019a) and in larger sociopolitical developments driven by market economies and neoliberal agendas, most educational institutions are probably quite far from implementing interspecies sustainability policies. The issue of food appears as particularly controversial and resistant to policy regulations. My university, University of Gothenburg, frequently claims to be at the forefront of sustainable development work. In Autumn 2018, a university employee initiated a debate in the university's internal newspaper, arguing for the implementation of a cross-faculty vegetarian food policy as part of the university's sustainability commitments (Gren 2018). The call was firmly rejected by the Vice-Chancellor, on the grounds that 'another policy' is not needed, and that it is the responsibility of deans, department heads, and individual working groups to take sustainable decisions (Wiberg 2018).

Institutional anxiety

The question I will pursue here is Deborah Britzman's (2003): What holds education back? 'The Sydney Declaration on Interspecies Sustainability' (Probyn-Rapsey et al. 2016) outlines a number of possible factors impeding cultural and institutional change around interspecies sustainability (especially animal consumption), including affective barriers to plant-based transition. My presentation will contribute to this discussion but from a slightly different angle, exploring institutional anxiety as an attempt to conceptualize what could be at stake. An assumption is that despite educational institutions' strong commitment to science and scientific evidence, responses to initiatives against animal-oppressive practices and conventions in education, even if motivated by sustainability concerns, are not necessarily based on rational considerations and decisions. Something else may be at work here.

I will turn to a Deleuze studies scholar, Andrew Culp, and Deborah Britzman's work in psychoanalysis and education, to begin to approach Britzman's question of what holds education back. Viewed through Culp's Deleuzian analysis (Culp and Dekeyser 2018), education is an apparatus of capture – a set of techniques associated with state power. Education holds itself back from igniting revolutionary instincts because it works in the area of chaos prevention and normalization of the present. (Ironically, chaos and disintegration of normality is likely to be part of a future of ecosystem collapse that could await us with or without educational consolidations of the status quo). The multiple stakeholder and market-driven requirements on education are part of a new landscape of control society (Deleuze 1992), accompanied by unspoken and subtle anxieties of the ever-present risk of not living up to the expectations haunting educational institutions and practitioners from every corner of society. In schizoanalytic terms (Buchanan 2013), education works in tandem with a set of socio-political machines, requiring education to conform, perform, and deliver certain pre-defined outcomes (cf. Ball 2003; Moran and Kendall 2009).

Turning to Deborah Britzman's (2003) psychoanalytic exploration of education, the question of what (anxieties) hold education back is not primarily embedded in a socio-political control regime, but rather within education itself, or, more specifically, in the difficulty of separating education from its fictions and its 'outside'. As Britzman notes;

> We find discussions of education and its nervous conditions in the most unusual places: in a spate of contemporary novels, where the university is the stage for betrayal and misunderstandings; in popular films, where heroic teachers, aliens, and teenagers vie

> for glory; in comic sketches that exaggerate the absurdity of school rituals; and, most intimately, in our dreams. (Britzman 2003, 6)

To these representations of education in popular culture as accommodating anxieties of all sorts, could be added affective investments in debates on what counts as education and miseducation; resistance to new knowledge; teaching as an 'impossible profession' guaranteed to produce unsatisfactory results; as well as those familiar tendencies of anxiety emerging from school experiences recognizable by all who are, or have been, school children: 'school phobias, running away, not being able to read, early stuttering, school pranks, and even cases of uncontrollable laughter.' (Britzman 2003, 16).

It is here somewhere, in the rather heterogenous nervous system of education, in a fuzzy zone between reality and phantasy, between uncertainty and authority, beyond deliberate planning and predictable outcomes, that anxiety intensifies, proliferates, and picks up speed. I suggest that institutional anxiety can be captured through this intersection of schizo- and psychoanalysis, and that it works as a defense against, and denial of the series of continuous breakdowns that may be momentary, but together constitute a 'permanent crisis' of education (Buchanan 2013). As we shall see, these breakdowns are sometimes produced by education's sociopolitical 'outside'; sometimes fuelled by education itself.

Human exceptionalism must be defended[2] (I): institutional anxiety in educational practice

A particular instantiation of institutional anxiety is evoked by threats to the anthropocentric infrastructure of education. Addressing a familiar setting of non-formal education – zoos and aquariums – Jason Hribal (2010) gives an account of historical cases when captive animals have violently resisted their confinement in establishments of the 'edutainment' industry of zoos, aquariums, and theme parks, often by killing their trainers. 'There is a long history to this struggle', remarks Hribal. 'Zoos and circuses live in fear of it and the historical changes that it can bring' (Hribal 2010, 152). The fear that Hribal speaks about is indeed an institutional fear, as 'incidents' of captive animals killing zoo staff (for whatever reason) is potentially disastrous to the reputation of any zoo establishment once the story hits the media, and it usually sets off a whole apparatus of crisis management and public communication strategies.

The institutional anxiety of interest in this paper is, however, of a less spectacular and more imperceptible kind. Rather than a straightforward narrative of 'nature strikes back' revenge, the anxiety in the schools and universities I have encountered is rather related to what Cary Wolfe (2003) has called 'the question of the animal'. As Jennifer Howard (2009) puts it in *The Chronicle of Higher Education*, '/ ... /[animal studies] scholars want to break down the categories and distinctions that have defined how we think about our relationship to everything that is not us. Some of them see it as nothing less than a revolution in how to think and how to live'. And, we could add, 'how to educate'. Thus, the effects from opening education to 'the question of the animal' may have a revolutionary potential not unlike the revolutionary potential of Hribal's (2010) violently resisting animals in zoos, because it disturbs prevailing anthropocentric mindsets and conventions in educational practice and theory. During my fieldwork in an upper secondary school with an animal caretaker programme, I interviewed an assistant principal, asking whether he thought that talented applicants to the school might be turned off by the information on the school's website, saying that students are expected to use animals for research purposes. He admitted that he could see a risk that some idealist-minded student would apply to the programme in order to engage in releasing (farmed) minks in the future. 'But then', he added, 'they surely get disappointed', because 'that is not what this programme is about' (*nervous laughter*), 'I mean it's not that we try to preach that we should engage in fur farming, it's not that, but ...' (Pedersen 2019a, 135). Surely there is a hint of anxiety in this school leader's response to the risk of attracting nonconforming animal activist students to the school and what this could eventually lead to.

In another interview, this time with an animal rights activist, she told me about her experiences from upper secondary school. While her activist engagement with the school's local Amnesty group passed by unnoticed, the principal reacted strongly to her displaying posters and T-shirts on the school premises with vegetarian and vegan messages – 'Choose life, choose vegetarian' – viewing it as an ostensible offence to other students. The principal asked her to take off her T-shirt and refrain from wearing it, but the student refused (Pedersen 2019a). This could be understood as another example of institutional anxiety – a fear of what parents and students may say if the school's unarticulated meat norm (Gålmark 2005) – a part of its anthropocentric infrastructure – is disturbed. Non-conforming, it seems, is a constant source of anxiety to educational institutions' investments in chaos prevention and normalization of the present (Culp and Dekeyser 2018), whether the non-conforming act consists of harmless school pranks and sudden outbursts of uncontrollable laughter (Britzman 2003), or politically driven expressions of animal rights messages in schools.

When school students initiate acts of resistance, although they clearly perform as norm-breakers, they also adhere to a rather conventional protocol around the power/resistance nexus in education, assuming that revolutions are primarily the arena of young people striving for influence over their daily lives as well as over social orders on a global scale. Youth resistance usually sets off an apparatus of control measures orchestrated from 'above', that is, from the authority collective of school leaders, teachers and principals. But what happens when change is driven by adult 'insiders'; for instance, when university teachers confront their own colleagues with requests for deconstruction of the anthropocentric infrastructure of their workplace? A few years ago I initiated a pedagogical development project in my department, with the purpose to introduce critical animal pedagogies as part of sustainable development in education. The project, involving critical analyses of animal exploitation by challenging educational anthropocentrism and speciesism, gained support from the department leadership and a small working group was formed (Pedersen, Håkansson, and Wals 2019). When it was time to implement the new course module however, there was fierce disapproval from our teacher colleagues. When they realised that we had included undercover footage from a pig production site and slaughterhouse as a learning resource in the module, all hell broke loose. One science teacher colleague even conducted a minor research investigation around the film, produced by an animal rights NGO, and presented evidence he found on the web indicating that parts of the film were not 100% authentic but actually manipulated in certain ways. This resulted in a series of breakdowns: the controversial video was taken out of the course module, and even led to a suggestion, supported by the Director of graduate studies, to take out the entire critical animal pedagogies module altogether.

When attempting to understand the twist and turns around this conflict, a colleague who was part of the working group concluded that our initial pedagogical idea of 'becoming uncomfortable' together with our students by asking them to critically question established norms of human–animal relations in society and education, became curiously enacted in the development of the module itself (Pedersen, Håkansson, and Wals 2019). Surely the discomfort surfacing during this process can also be understood as institutional anxiety: Anxiety evoked through affective investments in the debate on what counts as education and miseducation ('the pig slaughter video is biased'; 'critical animal pedagogies is a mission rather than education'; 'education needs to be objective and show both sides of an argument', cf. Flynn 2003); anxiety that such a marginal topic as human-animal relations would occupy a curricular space that could be used for more important matters; and anxiety emerging from the difficulty of separating education from its 'outside' (an animal rights NGO) (cf. Britzman 2003). I argue that institutional anxiety, in this case, gained particular force for two major reasons: First, because our course module was a bold challenge to the human exceptionalism underpinning the anthropocentric infrastructure of our teacher education department; and second, because the initiative was driven by teacher colleagues and staff – 'insiders' – rather than by students. The difficulty of separating education's 'outside' from its 'inside' in Deborah Britzman's (2003) psychoanalytic account is accentuated because there is no distinct boundary between them in the first place; rather, with Deleuze and Guattari ([1980] 2004), the interior and the exterior are co-constitutive, constantly folding in on each other, existing trough their exchanges. Thus, axiomatic human exceptionalism is an intrinsic part of education, but so is its countermovements and resistance.

Non-conformity and its associated ever-present risk of credibility loss in the eyes of stakeholders, but also fragility of the unconscious internal bonds that keep us together in education, seem to be what holds education back. If we leave educational practice for a moment and turn our attention to educational research, we will find similar tendencies at work.

Human exceptionalism must be defended (II): institutional anxiety in educational research

In his review essay, 'Listen, Ecological Marxist! (Yes, I Said *Animals!*)', John Sanbonmatsu (2005) critiques the anthropocentric foundations of Marxism, claiming that Karl Marx, like most other Socialists of his time, never questioned human domination of other animals. Marx, writes Sanbonmatsu, 'tried hard to distance his erstwhile scientific theories from all "bourgeois" claims of sentiment, including moral concern for animal welfare (which he viewed with undisguised contempt)' (113). Sanbonmatsu recommends contemporary Socialists still inclined to see animal rights as a fringe issue to take a lesson or two from analytic moral philosophers, such as Peter Singer and Tom Regan. Embedded in Sanbonmatsu's critique is not only an explicit request for a less anthropocentric critical theory, but also an implicit hint at an anxiety for 'the question of the animal' (Wolfe 2003) in leftist scholarship.

With some notable exceptions (Oakley 2019; Russell and Spannring 2019; Spannring 2017), reading current contributions to the research debate in Education for Sustainable Development (ESD) provokes a paraphrasing of Sanbonmatsu: 'Listen, Sustainable Development Educators! (Yes, I Said *Animals!*)', as unquestioned anthropocentrism appears as the baseline of the bulk of ESD research. However, this has not always been the case. Helen Kopnina (2012) describes a radical change of focus that took place with the shift from Environmental Education (EE) to Education for Sustainable Development (ESD). While earlier forms of EE were concerned with issues of environmental ethics, ecological justice and relations between humans, nature, and other species, Kopnina notes that with the turn to ESD, environmental protection was downplayed and social issues, 'which may or may not be related to environment' (701), prioritised. In ESD discourse, human rights are elevated, whereas the rights of other species are rarely addressed, and nonhuman animals are forced into generic categories as ecosystem representatives, biodiversity problems, and natural resources, their value defined instrumentally in relation to the extent to which they promote (or not) *human* well-being. The 'social sustainability' tendency in ESD has, it seems, driven interspecies relations to the margins, affirming an anthropocentric bias at a point in time when large-scale animal exploitation has been identified as a catalyst of innumerable environmental problems and a serious threat to the life-supporting ecosystems on which we all depend.

While Kopnina appeals to the empathy of individual ESD scholars to rethink the anthropocentric bias in ESD, I suggest that a re-definition of 'the social' in social sustainability to include human-animal relations is necessary, which would put ESD in closer conversation with developments in sociology (cf. Arluke and Sanders 1996) and sustainability studies (cf. Hiedanpää, Jokinen, and Jokinen 2012; Pöllänen and Osika 2018).

Institutional anxiety regarding the question of the animal in education research does not always take form as blatant marginalisation, disregard, or resistance. It may also, somewhat paradoxically, emerge as 'enlightened distance' to anthropocentrism (Pedersen 2019b). Rather than anxiety of the question of the animal, in this recent scholarship, we find efforts to bring human-animal relations *in* as integrated dimensions of ESD research which, at first sight, makes a promising move toward the inclusion of nonhuman animals in notions of the 'social' in social sustainability. A closer reading of these publications, however, reveals an animals-*for-us* position (Pedersen 2019b; cf. Wallin 2014) in solidarity with human-centered liberal pluralism masquerading as critique (Lindgren and Öhman 2019) or even in solidarity with the animal industry itself (Bruckner and Kowasch 2019), rather than solidarity with the marginalised group these authors claim to integrate (i.e., the animals). Purporting not only an 'enlightened distance' to anthropocentrism, but also a distance to what the authors perceive as 'moralizing' about the situation of animals in society, I suggest some element of institutional anxiety could be at work in these articles. This is an anxiety haunting scholars, myself included, who are openly advocating for nonhuman animals. It is associated with a risk of credibility loss, accompanied by the ever-present threat that one's status and career prospects (including possibilities for research funding and promotion) in academia may be jeopardized, due to one's solidarity with animals and animal rights social movements. Omnicidal times or not; human exceptionalism must be defended, and again, anxiety is at work in the folding in of uncomfortable exteriority (animal liberation advocacy) in the anthropocentric infrastructure of academia. The question then becomes, under what conditions can anxiety be a productive force that not only holds education back, but actually pushes education forward (Britzman and Dippo 2000)?

Educational resistance and change in omnicidal times

Erin Manning (2016) argues that the 'minor gesture', in contrast to grand manifestations or organised revolutions, is a force creating conditions for system-shifting changes. A student refusing to take off her T-shirt with an animal activist message in school, and a small teacher team developing a course module unsettling anthropocentric conventions in the teacher education department, can both be viewed as 'minor gestures' of resistance to the anthropocentric infrastructure of education. Both open possibilities for being and acting differently in education. Both are also acts of desire, permeated by passionate engagement, a stubborn belief that change is still possible and that education somehow *can* become a venue for enacting a different world. Desire, in Deleuze and Guattari, is a life-force; it can be put to work against omnicidal tendencies, and when condensed to a four-word T-shirt print worn by a secondary school student ('Choose life, choose vegetarian!'), this minor gesture becomes a desiring-machine (Deleuze and Guattari [1972] 2009):

> Desire is revolutionary by nature because it builds desiring-machines which, when they are inserted into the social field, are capable of derailing something, displacing the social fabric. (Deleuze 2004, 233)

Although T-shirt wearing and course module development as minor gestures of resistance in education may, as we have seen, stir up a great deal of conflict, they are in themselves no guarantee for change; their effects may indeed not change anything at all. What they do is that they point to the possible; the potential of education to exceed predefined positions presented to students (Pedersen 2019a), or, in Manning's (2016) words, 'what else life could be' (8). It would indeed be hopeful if these acts of resistance could release a multiplicity of desiring-machines within education, capable of (speaking with Deleuze) 'derailing' its anthropocentric infrastructure. However, Andrew Culp warns us (drawing on The Invisible Committee) that change is unlikely to be achieved from *within* education, as the insider position suffers from a 'curse of symmetry' occurring when we constitute ourselves on the same plane as the institution we seek to change. Rather, we need escape routes, carving out spaces for partisan knowledge aimed at coalition-building (Culp and Dekeyser 2018), refusing the normalizing language of an education seeking to colonize our future with models of the present.

Although I believe we have to be attentive to the 'curse of symmetry', and fully endorse the necessity to carve out escape routes and tactics of coalition-building with groups outside academia (such as animal advocacy and environmental movements), it would be a mistake to contend that working for change *within* education is futile. This is not a naïve over-reliance on what Peim and Stock (2021) have called the belief in the 'mythical powers' of education to bring about change. As educators, we must never forget the importance of inspiring, supporting, and collaborating with a new generation of students. As educators, we are not only engaged in 'teaching', but also in modeling different ways of *being* as an academic; different ways of being and acting and occupying the academic space. What we do (or refrain from doing) in this space, matters. Therefore, we should not let institutional anxieties hold education back, and we should not let institutional anxieties become our own (as we do when we make reassurances that our radical initiatives are negotiable, harmless, and nothing really to be anxious about). Rather, let us de-territorialize our teacher positions by openly acknowledging and confronting anxiety around human exceptionalism in education. If veganism, for instance, turns out to be an anxiety-producing topic in education, let us push it one step further beyond our comfort zones and explore scenarios of human withdrawal from violent intervention with the world; an absolute cessation of colonization, exploitation and omnicide (MacCormack 2013, 2020). Such a de-territorialization in education may offer impetus for action as well as mental preparation for a future that may be ours regardless of what happens (or not) in education. It may make present human exceptionalist-education increasingly unintelligible and intolerable – just like the world at large under omnicidal conditions.

Conclusion

It may appear that there are plenty of incentives for education to change. The threat of a disintegrating human civilization on a largely inhabitable earth is one of them, and could sooner or later make education in its present manifestations alienated even to itself (Wallin 2016). This existential dilemma (or drama) for education has been concisely summarized by Greta Thunberg's *Fridays for Future* movement: What is the point of educating for a future that may not exist? What makes Thunberg's straightforward question so startling is that under current planetary conditions, when collapsing ecosystems may become unable to support life as we know it, the very questions we ask about education must themselves change.

This paper has taken on board Deborah Britzman's (2003) question: What holds education back? I have explored the concept of institutional anxiety as both a blockage holding education back, as well as a potential force for change that may push education forward, especially regarding the anthropocentric infrastructure of education that is becoming increasingly obsolete in omnicidal times. The cases brought up from my fieldwork, however anecdotal and disparate, may serve as illustrating examples of how even minor disturbances to human exceptionalist assumptions may trigger quite strong reactions in education. Further research could explore more in-depth how education emerges in response to a multiplicity of anxiety-producing forces; however, the provisional analyses in this paper simply suggest that anxiety appears to be embedded as *one* dimension of the complexity of these responses. To put it differently, I do not claim institutional anxiety to be an explanation model, nor that this is the way we should look at education in order to understand and transform it. Rather, I see this concept as opening new modes of thought and spaces of action in education, as well as a tool to explore together with our students. Institutional anxiety may work as a consolidating force, but as a tool for analysis and impetus for action, it also troubles the assumption that education is immutable to change, and may create new conditions for being and acting in academia in omnicidal times.

Notes

1. Drawing on Sigmund Freud, Donna J. Haraway (2008, 11–12) defines human exceptionalism as a Western fantasy by which the self-centered human subject tries to hold panic at bay; a panic connected to three great historical wounds to the primary narcissism of the human subject: The Copernican wound, the Darwinian wound, and the Freudian wound; all three of which removed 'Man' from his superior position as the centre of the universe and above all other animals. Haraway describes these historical wounds as 'decentering cuts' inflicted by science.
2. The subheading paraphrases Michel Foucault ([1997] 2003)*'Society Must Be Defended'. Lectures at the Collège de France 1975–1976*. New York: Picador.

Disclosure statement

No potential conflict of interest was reported by the author.

ORCID

Helena Pedersen http://orcid.org/0000-0002-4693-3258

References

Arluke, A., and C. R. Sanders. 1996. *Regarding Animals*. Philadelphia: Temple University Press.

Ball, S. 2003. "The Teacher's Soul and the Terrors of Performativity." *Journal of Education Policy* 18 (2): 215–228. doi:10.1080/0268093022000043065.

Best, S. 2009. "The Rise of Critical Animal Studies: Putting Theory into Action and Animal Liberation into Higher Education." *Journal for Critical Animal Studies VII* (1): 9–52.

Britzman, D. P. 2003. *After-Education: Anna Freud, Melanie Klein, and Psychoanalytic Histories of Learning*. Albany: State University of New York Press.

Britzman, D. P., and D. Dippo. 2000. "On the Future of Awful Thoughts in Teacher Education." *Teaching Education* 11 (1): 31–37. doi:10.1080/10476210050020345.

Bruckner, H. K., and M. Kowasch. 2019. "Moralizing Meat Consumption: Bringing Food and Feeling into Education for Sustainable Development." *Policy Futures in Education* 17 (7): 785–804. doi:10.1177/1478210318776173.

Buchanan, I. 2013. "Deleuze and Guattari's Schizoanalysis: An Analytic for Education Assemblages?" Symposium CFP for Australian Association for Research in Education Conference 2013, Adelaide.

Celermajer, D. 2020. "Omnicide: Who Is Responsible for the Gravest of All Crimes?" *ABC Religion & Ethics*, January 3. https://www.abc.net.au/religion/danielle-celermajer-omnicide-gravest-of-all-crimes/11838534

Chakrabarty, D. 2009. "The Climate of History: Four Theses." *Critical Inquiry* 35 (2): 197–222. doi:10.1086/596640.

Cole, M., and K. Stewart. 2014. *Our Children and Other Animals: The Cultural Construction of Human-Animal Relations in Childhood*. Farnham: Ashgate.

Corman, L., and T. Vandrovcová. 2014. "Radical Humility: Toward a More Holistic Critical Animal Studies Pedagogy." In *Defining Critical Animal Studies: An Intersectional Social Justice Approach for Liberation*, edited by A. J. Nocella II, J. Sorenson, K. Socha, and A. Matsuoka, 135–157. New York: Peter Lang.

Culp, A., and T. Dekeyser. 2018. "On Giving up on This World: Andrew Culp, Interviewed by Thomas Dekeyser." *Society and Space*, February 27. https://www.societyandspace.org/articles/on-giving-up-on-this-world-andrew-culp-interviewed-by-thomas-dekeyser

Deleuze, G. 1992. "Postscript on the Societies of Control." *October* 59: 3–7.

Deleuze, G. 2004. *Desert Islands and Other Texts 1953-1974*, edited by David Lapoujade; Translated by Michael Taormina. Los Angeles: Semiotext[e].

Deleuze, G., and F. Guattari. [1972] 2009. *Anti-Oedipus: Capitalism and Schizophrenia*. Translated by R. Hurley, M. Seem, and H. R. Lane. London: Penguin Books.

Deleuze, G., and F. Guattari. [1980] 2004. *A Thousand Plateaus: Capitalism and Schizophrenia*. Translated by Brian Massumi. London: Continuum.

Foucault, M. [1997] 2003. "Society Must Be Defended". Lectures at the Collège de France 1975-1976 *Translated by David Macey. New York: Picador.*

Flynn, C. P. 2003. "A Course Is A Course, of Course, of Course (Unless It's an Animals and Society Course): Challenging Boundaries in Academia." *International Journal of Sociology and Social Policy* 23 (3): 94–108. doi:10.1108/01443330310790273.

Gålmark, L. 2005. *Skönheter och odjur: En feministisk kritik av djur-människa-relationen*. Göteborg och Stockholm: Makadam.

Gren, J. 2018. "En vegetarisk matpolicy sätter GU på kartan!" *GU-Journalen* 4: 40–41.

Gunnarsson Dinker, K., and H. Pedersen. 2016. "Critical Animal Pedagogies: Re-Learning Our Relations with Animal Others." In *The Palgrave International Handbook of Alternative Education*, edited by H. E. Lees, and N. Noddings, 415–430. London: Palgrave Macmillan.

Haraway, D. J. 2008. *When Species Meet*. Minneapolis and London: University of Minnesota Press.

Hiedanpää, J., A. Jokinen, and P. Jokinen. 2012. "Making Sense of the Social: Human-Nonhuman Constellations and the Wicked Road to Sustainability." *Sustainability: Science, Practice, & Policy* 8 (1): 40–49.

Howard, J. 2009. "Creature Consciousness: Animal studies tests the boundary between human and animal–and between academic and advocate." *The Chronicle of Higher Education*; Washington. October 18 http://chronicle.com/article/Creature-Consciousness/48804/

Hribal, J. 2010. *Fear of the Animal Planet: The Hidden History of Animal Resistance*. Oakland: AK Press.

Kahn, R. 2003. "Towards Ecopedagogy: Weaving a Broad-based Pedagogy of Liberation for Animals, Nature, and the Oppressed People of the Earth." *Journal for Critical Animal Studies* 1 (1): 36–53.

Kahn, R., and B. Humes. 2009. "Marching Out from Ultima Thule: Critical Counterstories of Emancipatory Educators Working at the Intersection of Human Rights, Animal Rights, and Planetary Sustainability." *Canadian Journal of Environmental Education* 14: 179–195.

Kopnina, H. 2012. "Education for Sustainable Development (ESD): The Turn Away from 'Environment' in Environmental Education?" *Environmental Education Research* 18 (5): 699–717. doi:10.1080/13504622.2012.658028.

Lindgren, N., and J. Öhman. 2019. "A Posthuman Approach to Human-animal Relationships: Advocating Critical Pluralism." *Environmental Education Research* 25 (8): 1200–1215. doi:10.1080/13504622.2018.1450848.

MacCormack, P. 2013. "Gracious Pedagogy." *Journal of Curriculum and Pedagogy* 10 (1): 13–17. doi:10.1080/15505170.2013.789994.

MacCormack, P. 2020. *The Ahuman Manifesto*. London and New York: Bloomsbury Academic.

Malm, A., and A. Hornborg. 2014. "The Geology of Mankind? A Critique of the Anthropocene Narrative." *The Anthropocene Review* 1 (1): 62–69. doi:10.1177/2053019613516291.

Manning, E. 2016. The Minor Gesture. Durham and London: Duke University Press.

Moran, P., and A. Kendall. 2009. "Baudrillard and the End of Education." *International Journal of Research & Method in Education* 32 (3): 327–335.

Nocella II, A. J.,C. Drew, S. K. George, J. Lupinacci, I. Purdy, and J. L. Schatz, eds. 2019. *Education for Total Liberation: Critical Animal Pedagogy and Teaching against Speciesism*. New York: Peter Lang.

Oakley, J. 2019. "What Can an Animal Liberation Perspective Contribute to Environmental Education?" In *Animals in Environmental Education: Interdisciplinary Approaches to Curriculum and Pedagogy*, edited by T. Lloro-Bidart and V. S. Banschbach, 19–34. Cham: Palgrave Macmillan.

Pedersen, H., J. Håkansson, and A. E. J. Wals. 2019. "Introducing Critical Animal Pedagogies in Higher Education." In *Vetenskaplighet i högre utbildning: Erfarenheter från lärarutbildning*, edited by O. Franck, 315–334. Lund: Studentlitteratur.

Pedersen, H. 2019a. *Schizoanalysis and Animal Science Education*. London and New York: Bloomsbury Academic.

Pedersen, H. 2019b. "The Contested Space of Animals in Education: A Response to the "Animal Turn" in Education for Sustainable Development." *Education Sciences* 9 (3): 211. doi:10.3390/educsci9030211.

Peim, N., and N. Stock. 2021. "Education after the End of the World. How Can Education Be Viewed as a Hyperobject?" *Educational Philosophy and Theory*. doi:10.1080/00131857.2021.1882999.

Pöllänen, E., and W. Osika. 2018. "Vi och 'de andra' – Social hållbarhet kopplat till hur i känner, tänker och agerar i relation till andra djur." *Socialmedicinsk Tidskrift* 2: 132–140.

Probyn-Rapsey, F., S. Donaldson, G. Ioannides, T. Lea, K. Marsh, A. Neimanis, A. Potts et al. 2016. "A Sustainable Campus: The Sydney Declaration on Interspecies Sustainability." *Animal Studies Journal* 5 (1): 110–151.

Russell, C., and R. Spannring. 2019. "So What for Other Animals? Environmental Education Research after the Animal Turn." *Environmental Education Research* 25 (8): 1137–1142. doi:10.1080/13504622.2019.1687639.

Sanbonmatsu, J. 2005. "Listen, Ecological Marxist! (Yes, I Said *Animals!*)." *Capitalism Nature Socialism* 16 (2): 107–114. doi:10.1080/10455750500108385.

Spannring, R. 2017. "Animals in Environmental Education Research." *Environmental Education Research* 23 (1): 63–74. doi:10.1080/13504622.2016.1188058.

Wallin, J. 2014. "Dark Pedagogy." In *The Animal Catalyst: Towards Ahuman Theory*, edited by P. MacCormack, 145–162. London: Bloomsbury.

Wallin, J. 2016. "Pedagogy at the Brink of the Post-anthropocene." *Educational Philosophy and Theory* 49 (11): 1099–1111. doi:10.1080/00131857.2016.1163246.

Wiberg, E. 2018. "Det görs redan mycket idag." *GU-Journalen* 4: 41.

Wolfe, C. 2003. *Zoontologies: The Question of the Animal*. Minneapolis and London: University of Minnesota Press.

Wood, L. 2020. "Have More than a Billion Animals Perished Nationwide This Bushfire Season? Here are the Facts." *RMIT ABC Fact Check*, February 3. https://www.abc.net.au/news/2020-01-31/fact-check-have-bushfires-killed-more-than-a-billion-animals/11912538

Animal advocacy, fear and loathing in academia: A response to Helena Pedersen

Kai Horsthemke

ABSTRACT
Helena Pedersen's powerful keynote address poses the question: What prevents education from becoming a transformative force in times of 'omnicide', that is, 'the annihilation of everything'? She locates at least part of the response in 'institutional anxiety', which constitutes a (social-) psychological barrier to radical change. In particular, she discusses anxiety related to the moral standing of non-human animals as a threat to human exceptionalism in educational practice and research. Institutional anxiety, as I show in my discussion of a recent manifestation at a university in South Africa, also occurs in post-liberation societies, when 'university teachers confront' or consider confronting 'their own colleagues with requests for deconstruction of the anthropocentric infrastructure of their own workplace' (Pedersen), colleagues they know to have been historically marginalised or disenfranchised. However, as I hope to make clear in my response, there are some ways of confrontation that are less divisive than others.

When I invited Helena Pedersen to deliver a keynote address at INPE 2020, I did so because my familiarity with her work gave rise to certain expectations – expectations that, I am happy to say, have been fulfilled . . . and then some! Her essay gives us not only philosophically compelling arguments and generative pedagogical ideas to think about, but also an insight into her deeply personal commitments.

So, rather than embark on a 'critical response' and/or some conceptual nitpicking (which would be highly artificial), I would like to dwell on an incident that illustrates *both* the complexity of animal (and environmental) advocacy within academia *and* institutional anxiety. Apart from the instantiations mentioned by Pedersen, institutional anxiety may also have a cultural dimension.

In September 2015, a proposal was submitted to the Humanities Faculty at the University of Cape Town, South Africa (Thamm 2016; Glover 2017; Cordeiro-Rodrigues 2020). Written by four academics, all of them faculty members, it requested that, given the suffering and killing involved in producing food from animals, no animal products be served at any Humanities Faculty meeting or event. The proposal produced a substantial outcry among other faculty, who saw this advocacy of veganism, like all criticisms of certain cultural practices concerning animals, as culturally insensitive and disrespectful, racist and/or 'white normative' (Cordeiro-Rodrigues 2020). Faculty members promoting animal-related issues, 'it was suggested, were deprioritising key issues of race and colonialism' (Glover 2017, 176–177; Thamm 2016). The proposal was supported by 8 faculty members, rejected (not supported) by 50, with 25 abstentions (Thamm 2016).

Like the charge of racism, the notion of 'white normativity' (employed as interchangeable with 'normative whiteness'; see Cordeiro-Rodrigues 2020) is somewhat misleading and, at best, of limited usefulness. In this particular instance, faculty members initiating the 'animals off the menu' proposal in question may have all been white. But veganism is not a 'white' norm. Apart from being inherently opposed to all injustice, discrimination and arbitrary hierarchical power structures along racial, gender, cultural and species lines, it is a lifestyle also advocated by the likes of Gandhi, the Dalai Lama and others – *vide* Jainism and other Eastern traditions. Moreover, it is at least *conceivable* that the proposal could have had advocates who were 'of colour' (to use a popular, albeit somewhat unreflective phrase). The Clifton Beach sheep slaughter protests[1] in Cape Town, South Africa, for example, included many who were not white – in which case the conflict is arguably not one sparked by 'white normativity' but, rather, by a failure to communicate principled moral opposition to all oppression, abuse and exploitation. Are the respective cases against clitoridectomy, female genital excision, and against 'virginity testing' and related practices[2] also examples of 'white normativity'? What about the prosecution of poachers on the African continent?

Pedersen's reference to the 'anxiety haunting scholars openly advocating for non-human animals' being 'associated with a risk of credibility loss, accompanied by the ever-present threat that one's status and career prospects (including possibilities for research funding and promotion) in academia may be jeopardised', now receives an additional dimension. White South African academics may be increasingly reluctant to publicise or act out their 'solidarity with animals and animal rights social movements', for fear of being labelled 'racists' or pronounced guilty of cultural and moral hegemonism.

Perhaps an elegant solution would have been to point to elements within African moral thought that would arguably favour resolving this issue in one way rather than another, first and foremost the strong relational element contained in moral thought and practice on the African continent. *Ukama* and *ohanife* are Shona and Igbo terms in Zimbabwe and Nigeria, respectively, for relations/relationality. Like the principle of African humanism, 'I am because we are', they signal kinship beyond the species boundary and our essential relatedness and mutual dependence (see Murove 2009; Horsthemke 2015; Chimakonam 2018; Cordeiro-

Rodrigues 2020). This understanding is also captured in the Egyptian notion of *maat*, the Cree concept *wahkohtowin*, and the Lakota phrase *mitakuye oyasin*: 'all my relations', or 'we are all related'. My hunch is that this strategy is a powerful one that, coupled with the (wholly plausible) argument that the initial UCT proposal had an alienating effect, might have evoked greater sympathy, open-mindedness and willingness to change on the part of black faculty.

After all, an Afro-communitarian proposal for animal justice would also be in keeping with the much-touted Africanisation of educational institutions.

Institutional anxiety, then, also occurs in post-liberation societies, when 'university teachers confront' or consider confronting 'their own colleagues with requests for deconstruction of the anthropocentric infrastructure of their own workplace' (Pedersen, in this Issue), colleagues they know to have been historically marginalised or disenfranchised. However, as I hope to have made clear here, there are some ways of confrontation that are less divisive than others.

Notes

1. https://www.timeslive.co.za/news/south-africa/2018-12-28-sheep-slaughtered-on-clifton-beach-as-animal- rights-activists-protest/ Clifton has four separate beaches on the Atlantic Ocean and has always been the preserve of the affluent, mainly white residents living along the steep cliffs of the affluent suburb. In 2018, a group of activists allied to the '#FeesMustFall' and '#RhodesMustFall' movements at South African universities (see Horsthemke 2017) argued that rich whites had, for too long, polluted the beaches of Cape Town that now needed to be cleansed with a typically African ritual, involving the sacrificial slaughtering of a sheep or goat. The group mobilised hundreds of supporters. They arrived with a sheep to perform this ritual on the beach, much to the dismay of local residents and of animal rights groups, who in turn mobilised support. The two groups clashed on Clifton beach #4. However, the sheep was killed and her blood splattered around the beach, supposedly cleansing it of bad spirits and appeasing the ancestors. Opponents of this ritual were, of course, deeply alarmed by this spectacle and called on the city of Cape Town to re-establish some form of law and order on this normally quiet beach. This in turn fuelled the fires of the former anti-apartheid machinery who perceived the incident and the reactions it provoked as an opportunity to put one over on the privileged whites living in Clifton.
2. 'Virginity testing' in girls, also known as *inkciyo* or *ukuhlolwa kwezintombi* in the Nguni language group of southern Africa, is a ritual in which girls undergo routine hymen inspection by designated female community elders, a practice that 'seeks to achieve the goal of purity in the context of the spread of HIV/AIDS' (Ntuli 2002, 62), while *amaqhikiza* refers to a type of mentorship programme among older and younger girls 'to ensure sexual abstinence' until the latter are 'ready to take full control of their affairs' (ibid., 61). What arguably makes practices like *inkciyo* or *ukuhlolwa kwezintombi* objectionable is not only that similar testing is not applicable (and consequently never done) with regard to boys, but also that it is women who are implicitly held responsible for the spread of HIV/AIDS (see Horsthemke 2021, 198, 209).

Disclosure statement

No potential conflict of interest was reported by the author.

References

Chimakonam, J. O. 2018. "*Ohanife*: An Account of the Ecosystem Based on the African Notion of Relationship." In *African Philosophy and Environmental Conservation*, edited by J. O. Chimakonam, 120–134. London and New York: Routledge.

Cordeiro-Rodrigues, L. 2020. "Racialization of Animal Advocacy in South Africa: A Case Study from the University of Cape Town." *Ethnicities*. https://journals.sagepub.com/doi/10.1177/1468796820946762#.

Glover, M. 2017. "Animal off the Menu: A Racist Proposal?" In *Animals, Race, and Multiculturalism*, edited by L. Cordeiro-Rodrigues and L. Mitchell, 175–199. Cham: Palgrave Macmillan.

Horsthemke, K. 2015. *Animals and African Ethics*. Basingstoke: Palgrave Macmillan.

Horsthemke, K. 2017. "'#factsmustfall'? – Education in a Post-truth, Post-truthful World." *Ethics and Education* 12 (3): 273–288. doi:10.1080/17449642.2017.1343620.

Horsthemke, K. 2021. *Indigenous Knowledge – Philosophical and Educational Considerations*. Lanham: Lexington Books.

Murove, M. F. 2009. "An African Environmental Ethic Based on the Concepts of *Ukama* and *Ubuntu*." In *African Ethics: An Anthology of Comparative and Applied Ethics*, edited by M. F. Murove, 315–331. Scottsville: University of KwaZulu-Natal Press.

Ntuli, P. P. 2002. "Indigenous Knowledge Systems and the African Renaissance: Laying a Foundation for the Creation of Counter-hegemonic Discourses." In *Indigenous Knowledge and the Integration of Knowledge Systems: Towards a Philosophy of Articulation*, edited by C. A. Odora Hoppers, 53–66. Claremont: New Africa Books.

Thamm, M. 2016. "Fanon Meets Biko Meets JM Coetzee as UCT Academic Row over Food Highlights Racial Fault Lines." *Daily Maverick*. https://www.dailymaverick.co.za/article/2016-07-13-fanon-meets-biko-meets-jm-coetzee-as-uct-academic-row-over-food-highlights-racial-fault-lines/.

Be the village: Exploring the ethics of having children

David Chang

ABSTRACT
The rapid increase in human population is one of the underlying factors driving the ecological crisis. Despite efforts on the part of educators to raise awareness of environmental issues, the ecological impact of a burgeoning population – and the ethical implications of having children – remains an unbroachable topic. Nevertheless, the increase in human numbers is central to questions of sustainability: How can a species expect to survive in a finite terrestrial environment without limits to its population?

Since most of the world's ecological impact can be traced to capitalist-industrial- consumer societies in over-developed nations, the middle and upper classes in rich countries must weigh the ecological consequences of their family-planning decisions. In this paper, I argue that educational programs that are concerned with environmental ethics should have students examine the assumptions and implications of having children. I consider the risks associated with this proposal and respond to a series of possible objections. This paper does not advocate coercive measures for population control, but rather enjoins a pedagogical responsibility to view having children as an act with ecological consequences, an act that must be subject to careful examination.

Outset

To be alive in the second decade of the 21st century and relatively aware of the current picture of planetary plight, is to exist in a state of alarm, a disquieting anxiety in a moment of unprecedented disruption. Previous to this era, few had to fathom the implications of the Anthropocene (Waters et al. 2016), a new geological epoch prompted by the rise of *Homo Sapiens*. Although the future remains uncertain, a series of indicators portend troubling times ahead: runaway climate change (Reuters 2019), mass extinction (Dasgupta and Ehrlich 2019), plastic pollution (Singh 2019; Carrington 2019), habitat destruction ('Living Planet Report' 2020), and the spread of zoonotic pathogens due to deforestation and land degradation (Sachs et al. 2020). This concatenation of catastrophes indicates an ecological crisis that grows more dire by the day.

By in large, governments and societies have been unable to muster a sufficient response to address the spectre of climate change.[1] The recent rise of climate movements and ambitious government policies are encouraging. However, carbon is merely a symptom of a pervasive social-economic complex operating on principles and assumptions that make ecological devastation all but inevitable. A radical examination of the current establishment must venture beyond carbon emissions *per se* and critique the motive forces that compel human development in its current form. Attempts to address the ecological crisis are bound to fail if we do not reform the *bedrock values* that compel human action, individual and collective. Insofar as the cultivation and reformation of values have fallen within the purview of education, a deeper understanding of the roots of the ecological crisis is not possible without the efforts of educators everywhere. Operating under the industrial-economic paradigm that prepares students for a life of production and consumption (Orr 2004), schools in the West have been ineffective in addressing the ecological crisis.

In this paper, I argue that environmental education should include a discussion of the ecological consequences of having children. With the current planetary situation in mind and drawing on available evidence, there is a case to be made that having fewer or no children is the more environmentally responsible action one can take while living in a modern capitalist-industrial economy. To substantiate my case, I first outline the salient perspectives that compose the landscape of ecological ethics and identify the entrenched values that lie at the root of the ecological crisis; namely, anthropocentric beliefs about the privileged status of human beings in relation to the ecological world. Drawing on the work of Curry (2011), I discuss the tension between anthropocentric and eco-centric ethics. While the former attempts to induce changes in behaviour while leaving human-centric assumptions and values undisturbed, the latter attempts to unsettle the ontological suppositions that assume the primacy of the human animal. Based on Curry's (2011) conceptual equation (I = PLOT), members of the middle and upper classes in over-developed nations carry the responsbility of examining the implications of having children. This subset of the global population tends to have the greatest ecological impact, and thus carries the onus of careful consideration in matters of family planning. Finally, I respond to a series of possible objections to my proposal before concluding this paper with a reflection on the challenges that educators face if eco-centric education is to gain wider purchase.

At the outset, I must be clear about what I am *not* arguing for. First, I am *not* arguing for government policies and legislation that impose restrictions on reproductive rights. The history of population control measures (such as the one-child policy in China and forced sterilization in India) have been convulsive to the social fabric and have resulted in abject misery. Instead, I advocate for reasonable, honest, fact-based discussions within educational settings about the ecological impacts of having children.

Second, I am *not* arguing for a reduction in population, or a reduction in rates of population growth, in developing countries. As I will make clear in the paper, because the bulk of the current ecological impact can be attributed to those living in rich countries of the global north, the onus falls on the members of these nations to examine their collective impact. Although millions are poised to enter the middle-class in developing economies, and will collectively exert additional strain on ecological resources already over-taxed (Wheary 2010; Lange and Meir 2009), in this paper I leave aside this segment of the global populace and address instead the living standards to which they aspire.

Third, I am *not* arguing for a kind of educational program that would force students to adopt a predetermined conclusion, a form of indoctrination that stigmatizes or prohibits having children. Instead, I argue that in societies where having children is considered a given norm of adulthood, there should be conversations that facilitate careful consideration in this elective act. This is an educational attempt to question the unquestioned norms, to acquaint students with the arguments pertinent to a decision, and to examine the pervasive values that circulate within a culture. While such an investigation entertains inconvenient facts and brooks sensitive discussions, it need not lead to moral directives that circumvent critical thinking and independent decision-making.

In presenting this case against having children, I am not endorsing a cynical view of humans as a planetary plague. Caution and concern should not be mistaken for malevolence and antipathy, and educators should guard against what Karen Litfin calls 'the misanthropic temptation' (2013, 153). Concern over growing numbers does not contravene the affirmation of human value; adherence to limit coheres with a caring concern for the welfare of existing populations, human and more-than-human. Further, although I present a case against having children, this paper should not be read as an argument against *parenthood*. In the face of the ecological crisis, educators face the cultural task of re-shaping notions of parenthood, such that kinship extends beyond sanguinity. For prospective parents, this means finding ways to play a positive role in the lives of children through their professions, volunteer work, or through communal relationships.

Anthropocentrism

Anthropocentrism can be literally understood as human-centeredness, a fundamental assumption of human primacy, the belief in the pre-eminent status of homo sapiens in a supposed hierarchy of species, the primary holders of intrinsic value. Patrick Curry provides the following definition: 'the unjustified privileging of human beings, as such, at the expense of other forms of life' (Curry

2011, 55). Under the anthropocentric world view, the planet – all its lands, waters, plants and animals – appears to serve human utility; the decimation of wildlands and waterways constitute no crime, for they are necessary to the advancement of a human enterprise. According to anthropocentrism, the continued expansion of economies is only a matter of course in a world where nature is given to human use.

In stark relief to anthropocentrism, ecocentrism is more concerned with the well-being of the ecological system as a whole. Humans are seen as a constituent of the planet and enjoy no privileged status that would unduly subject the ecological community to the service of narrow human interests. Curry writes: 'ecocentrism recognizes, as anthropocentrism does not, that human beings live in a more-than-human world' (2011, 59). Thomas Berry (2006) has asserted that 'in all our activities the Earth is primary, the human is derivative' (p. 43). Aldo Leopold's (1986) *land ethic* is often presented as the definitive statement on ecocentric philosophy: "a thing is right when it tends to preserve the integrity, stability and beauty of the biotic community. It is wrong when it tends otherwise" (1986, 262). The land ethic locates humans within a web of interdependent relationships, with the ecological base at the centre of ethical considerations. Human interests are not always at odds with the interests of the ecological community but, given the interdependent relationships in which humans find themselves, there is no such thing as a human interest that can be easily extricated from the well-being of the ecological community. Pragmatically, an ecocentric ethic demands that where short-term human interest conflicts with the good of the ecological community, the interests of the latter must prevail (Curry, 2011).

An ecocentric ethic calls into question the foundational premises on which many societies are built, for in asserting the primacy of the land, ecocentrism attempts to reform the very assumptions that inhere human projects, which often take for granted the degradation of ecosystems. Where human activity dramatically reduces the vitality of a biotic community and impairs its regenerative capacity, ecocentrism requires that human actions be subject to limits. Ecocentrism examines the roots of the ecological crisis and poses a challenge to pervasive ideas about human entitlement – among them, unbridled population growth.

The impact of having children

Human overpopulation is one of many serious problems with dire ecological ramifications. Convincing evidence indicate that rapidly growing population exert a deleterious effect on the ecological base by way of water and food consumption, waste disposal, industrial pollution and effluence, sewage and sanitation, habitat destruction, depletion of fish and wildlife stocks, and greenhouse gas emissions (Chiras 2014).

Further, a growing human population with the material affluence seen in developed countries magnifies ecological impact by several orders. An overall picture of ecological impact is captured by the following conceptual formula: I = PLOT, Impact as a function of Population, Lifestyle, Organization and Technology (Curry 2011). Population refers to the total number of individuals who comprise the human species; Lifestyle marks the material demands of a given population; Organization refers to the arrangements that constitute a social order; Technology refers to the tools that humans use to accomplish a variety of tasks. A country like Canada, with a relatively small population compared to its G8 peers, exerts a staggering ecological impact due to the lifestyle of its average citizen, who enjoys high standards of living as measured by their consumption of goods and services. On the organizational front, corporations shape vast sectors of the economy and exert significant influence on government policy. Canadian corporations also deploy advanced and powerful technologies, such as autonomous mining systems, which exploit resources with staggering efficiency (Leah H 2018). Each factor in the equation cannot be considered in isolation from the others. Given a large enough population, even moderate lifestyles can result in significant ecological impact; reductions in demand for resources from efficient technologies can be offset by unbridled population growth.

It takes four to five times more land, forest, water, and carbon to raise a child according to the average Canadian standard of living compared to raising a child in India (Ewing et al. 2010). In 2018, the average Canadian emitted twice the amount of carbon compared to the average Chinese, and eight times more than the average Indian (European Union 2019). These measurements only account for the direct use of fossil fuels; if we take into the account the indirect, systemic use of energy associated with each Canadian (energy for production and transportation of goods and services), the picture becomes more alarming. A child born into a middle-class family in a developed country is associated with 58.6 tonnes of carbon per year, an impact that far outweighs the carbon impact of driving a car at 2.4 tonnes of carbon per year (Wynes and Nicholas 2017). Of all the carbon-reduction measures a person can undertake, having fewer or no children, is far more effective than living car-free, eating a plant-based diet, foregoing air-travel, and switching to renewable energy combined (Wynes and Nicholas 2017).

Despite the glaring ecological problems posed by over-population, most discussions around environmental education do not acknowledge the ecological dimensions of population growth. To implicate humans, by the very fact of their existence, as an ecological threat does not sit well with established anthropocentric suppositions and deeply inscribed beliefs about humans as an unalloyed good. From an ecocentric perspective, we might ask: *good for whom?* Certainly not for the animals slaughtered for food and forests razed for monocrop agriculture. While we can affirm the intrinsic and irreducible value of human beings, it does not follow that the arrival of more individuals increases the aggregate value of a population (the more the better), or that their intrinsic value entitles humans to propagate without limit. Ecocentrism does not deny the intrinsic value of a human being, only that human value should be considered alongside other species and entities, whose function within local ecosystems constitute the integrity of the whole planet. The question of population growth, therefore, is the crux of the tension between anthropocentrism and ecocentrism. Human insistence on continued and unbridled propagation, despite consequences to biodiversity, marks the extent of our ecological ethics. Ecocentrism ends where humans are unable to adhere to limits.

If we count among the worthy aims of education the development of awareness of global issues – such as exploitative labour practices associated with fast fashion, and the environmental impact of meat consumption – then an examination of a culturally sanctioned act such as having children should be included in the curriculum. Critical review of family planning need not force students to arrive at a certain conclusion, just as a review of the environmental consequences of eating meat should not force students to become vegans. Insisting that students arrive at a pre-conceived conclusion is the purview of dogma. Instead, educators should help students bring thought to action, and awareness to behaviour. The choice to have children is no exception.

This educational position effectively scrutinizes common attitudes about family planning, and is bound to infringe upon cultural, religious and established social beliefs, inciting vigorous debate in its wake; we should hardly expect anything less given the entrenchment of anthropocentric values. Yet whatever difficulties that might arise, an educational effort to curb the rise in population remains preferable to the other methods of population control that we have seen in the twentieth century. The one-child policy in China and forced sterilization in India both engendered innumerable human-rights abuses, most of which were disproportionately borne by the poor. The history of population policy is a lamentable one indeed, and an educational attempt to apply ethical examination to population growth, however challenging, remains preferable to the blunt instruments of state policies seen elsewhere.

Objections and responses

The call to educators to address the issue of population growth is likely to elicit several reactions. Here, I provide a brief response to several possible objections.

A) *Efforts to curb population are misanthropic and problematize humans by viewing their existence in a negative light.*

It is not misanthropic to care deeply for the lives of those who already inhabit the planet, who must struggle with survival when ecological systems decline; nor is it misanthropic to try to circumvent foreseeable suffering for the existing millions by preventing more births. Furthermore, prospective parents must contemplate the implications for children who must live in a world of great environmental upheaval. Consider the melting of arctic permafrost, an event that will likely increase concentrations of methane in the atmosphere, thus triggering run-away warming (Reuters 2019; Watts 2020). Such evidence does not seal the future but provides disturbing indications about a pending maelstrom. Children born today can expect a radically different world than the one their parents grew up in; they cannot count on the relatively stable and benign planetary conditions that previous generations took for granted. A careful consideration of the world in which a child must live springs not from a disdain for humanity, but rather from a loving concern for children and the plight of millions.

Ironically, those who refuse to enter discussions about over-population may unwittingly condone large scale suffering, thus assigning a large portion of humanity to misery. With the arrival of each new generation comes the responsibility, on the part of families (and societies), to provide nourishing food, clothes, shelter, a clean environment, education, employment, and adequate health care. Refusal to provide such goods and services leads to abject poverty and suffering. On the other hand, the provision of these goods necessitates more ecological devastation by way of resource extraction, habitat destruction, species extinction, and the like. Those who condone preventable births are therefore caught in a dismal bind: to provide the goods required for populations to flourish means a commitment to further ecological damage; failure to provide these goods lead to massive deprivation (Curry 2011). In the end, the more misanthropic position may be that which fails to account for the myriad of problems that await if population grows unchecked.

In considering this proposal, those who already have children may find themselves on the defensive. They may feel that by presenting a case against having children, I inadvertently impugn parents' ethical choices, and paint their children, by the very fact of their existence, an environmental problem.

In response, I submit that in the ethics of population, there are *matters of deliberation* and *matters of fact*. Before a child exists, a couple is free to weigh the implications of a pending decision. No child exists to bear the consequences of their deliberation – all possibilities are on the table. However, ethical discussions shift when a child is born. The moral landscape changes; priorities change. The child is now a *matter of fact* and presents a moral obligation by virtue of their existence. The child is an end in themselves, a given fact of the world, and the parents' responsibility is to their healthy upbringing. My proposal to examine assumptions around having children is directed toward *matters of deliberation*, not toward matters of fact. Once a child is born, they are part of the human collective whose welfare needs to be safeguarded against burgeoning numbers. Parents who already have children should view my proposal as an argument against having *another* child rather than as a condemnation of their existing family.

B) *Reproductive rights are a non-negotiable freedom*

No right is absolute and above debate and contention. Likewise, no freedom is so sacrosanct as to elude a requisite responsibility. The freedom to have children must attend the responsibility to provide children with a nurturing environment that supports their physical and emotional development. Parents owe this responsibility not only to the children themselves but also to the society that shares the social sphere with each child. It is ethically untenable to uphold the reproductive rights of *child-bearing* without assuming the task of *child-rearing* as a basic responsibility. By extension, the consequences of over-population requires that people 'fully come to terms with the fact that a person's (biological) right to have children must be mediated by his or her (social) responsibility not to have too many' (Smail 1997). Because every person across the globe is affected by the ecological crisis to some degree, social responsibility can no longer be limited to one's immediate circle of relations, to one's provincial or national membership, but extend to the family of humanity in its entirety.

C) *The proposal is anti-educational, because, following the logic of the argument, we can suppose that one day there may not be any children left to teach.*

If by 'anti-educational' we mean that the proposal deprives teachers of pupils, the absence of whom renders teachers jobless, then the objection can be addressed immediately. Teachers are gainfully employed because there are children to teach, but children do not exist so that teachers can have jobs. To render children the clientele to the teachers' services is to confuse ends and means, a profound misapprehension of the function of education within society.

On the other hand, perhaps the objection reveals a deeper concern. An honest look at our current predicament seems to cast a shadow on the human species as a whole. This conclusion seems to imply that *children themselves are ecologically problematic*, a heinous thought that is difficult for educators to hold. Thus, a paradox emerges from contradictory propositions: Every child is an immeasurable gift with intrinsic worth and dignity, and serious consequences are in store when more children enter the population.

The educational and theoretical challenge lies in articulating a coherent pedagogy out of an apparently incoherent message: *children are wonderful, and we should be careful about having more of them.*

In order to stabilize this dissonant view of children, I point to the difference between the *actual* and the *potential*. Children who are alive, running in the school yard, are wonderful and beautiful; they deserve their teachers' devotion and the conditions that nourish a healthy life. However, those 'not yet born' do not exist except in the realm of adults' imagination. To forego having children, therefore, does not diminish one's regard for existing persons who live and breathe amongst us. Many educators advocate for a limit on the number of students in a teacher's charge in order to ensure adequate attention and support for each pupil. By lobbying against increased classroom enrolment, teachers are not denigrating the value of their existing students, but rather guarding against the potential erosion of learning conditions. Similarly, to decide against having children is to defend against further ecological damage. The very idea of sustainability involves adhering to short-term limits to ensure long-term continuance.

Careful reflection on hither side of decision-making can be a gesture of loving responsibility rather than an agenda that undermines the educational project.

Pedagogical challenges

The foregoing responses highlight problems associated with an anthropocentric refusal to curb population growth. However, the tougher challenge lies in the pedagogical task of critically examining the impact of having children from an ecocentric perspective. Teachers walk a delicate balance between helping students confront the ecological impact of human expansion on the one hand, and navigating the *ontological shock* (Todd 2000) of encountering their complicity within systems that exert environmental impact while nourishing students' nascent identities and their inchoate confidence. Since ecocentrism places the interests of the ecological community at the heart of ethical deliberation, students who undergo an ecocentric education cannot avoid an encounter with human hegemony and the myriad ways that human supremacy excludes more-than-human world from ethical consideration. Although ontological shocks are necessary to the retrenchment of anthropocentric notions, they need not eviscerate students' sense of self-worth, nor should they devastate children's faith in the human capacity to change and respond to challenge. Although the honest acknowledgement of benefits gained at the expense of the more-than-human world can subject students to an experience of ontological guilt, such guilt need not become a withering force that diminishes one's capacity to act and inquire. Rather, guilt signals a growing awareness that is imperative to moral action. Ecological guilt has the potential to spur the development of new consciousness – its unsettling valence effectively becomes a *salubrious pain* that continuously peels away scales of blindness while challenging the implicate order.

The examination of the ethics of having children, then, is a way for students to realize the extent of their ecological impact within the system they have inherited. This recognition opens the possibility for *moral susceptibility* (Todd 2000). A developing understanding of the impacts of having children marks one dimension of a growing awareness that composes a life of action shaped by an ecological consciousness, an ecocentric awareness that imbues all aspects of a worldview, family planning included.

Reproductive restraint requires alertness to global implications, and recalls the Kantian axiom: *Act only according to that maxim whereby you can, at the same time, will that it should become universal law* (Kant 1993). The ethics of having children, according to the categorical imperative, cannot be determined by its benefits to the individual alone, but rather by the larger implications should the act be made universal.

Kant's insight anticipates the current debate that unfolds from the ecological crisis: Because the biosphere shelters every member of the human race, adverse consequences of ecological damage are shared by everyone; legitimate ethical deliberation must account for the good of the planetary community.

However, the problem of population lies partly in the scale at which analysis and discussion are conducted. The ecological impact of overpopulation is clear and urgent at a global level, but the seriousness of the crisis does not register easily at the level of everyday life. *Population* is a cold, statistical and faceless abstraction that is powerless to evoke the affection and sentiments attached to babies and children. We find it hard to fathom the plight of anonymous millions when we are captivated by the vivacious laughter of a child. Nevertheless, the need for ethical examination around matters of family planning only makes sense when global consequences are thought to matter. For educators, the pedagogical challenge lies in pointing out the immediate, relevant, and apparent effects of population growth, how such developments affect landscapes urban, rural, and wild. Teachers are thus tasked with making the abstract tangible, and the global personal. The extrapolation of historical aerial photographs of cities, towns, and agricultural lands alongside statistical graphs of population growth can present a correlation between population growth and changes in land use.[2] Teachers can generate discussions around what students have noticed around population growth, what they have noticed in their own daily lives (traffic congestion, urban sprawl, bigger crowds at public events). These discussions can bring awareness to the incremental changes that students experience and link local development to global patterns.

To illustrate the conundrum of unbridled population growth within terms that students can grapple with, consider the following classroom activity, a variation of a 'fishbowl' activity, whereby a small group of students work through a problem in the presence of peers who serve as observers. In this activity, the observers watch four students standing around a delicious pie. The four students try to share this pie fairly among them. The class watches as these students negotiate a fair division. Outside of the classroom, a several more students await. Just as the four students come to an agreement about how to share the pie, one student walks in and demands her share of the pie, thus overturning the newly forged agreement between the four. The students must negotiate again, and just as they reach a new agreement, another student walks in and demands a share of the pie. The negotiation begins again, and what tentative agreement the party comes to is annulled by yet another student who enters the room and demands a fair share of the pie. The dynamics of the interaction, the arguments issued in each round of negotiation, serve as indicators of the contests that occur when the supply of pie cannot meet the increase in demand.[3] In this activity, the observers are tasked with analysis of the problem and provide comments on how to resolve the issue, as well as the conundrums that arise when issues of fairness come up against unchecked population growth. The activity instantiates a way in which thoughtful consideration of family-planning need not proceed in the form of injunctions and pre-determined conclusions issued by a teacher, but rather through discussion and exploration that leave the students to grapple with ethical dilemmas inherent in certain situations.

A review of the ethics of having children, with its requirement for rational reflection, might be viewed as a campaign against intuitive longing, cultural tradition and social norms. For many, there is an emotional and often physical longing to have children of one's own, a longing that can persist despite the recognition of our ecological predicament. However, there are many ways of experiencing *parenthood* without bearing one's own biological children. The proposition is not hostile to parenthood, but rather encourages kinship of many kinds, human and more-than-human. If it takes a village to raise a child, one partakes in parenthood by *being* the village for existing children. Prospective parents can consider adopting children as a first option, and/or share care-giving duties with other parents within their familial and social network. This means establishing a nurturing relationship with the children in their lives, taking on the responsibility of being a mentor and guide to children and young people. Teaching, coaching, tutoring, can serve alternate forms of connection that offer the warmth of relationship without adding to the ecological pressures of a growing population. Following Donna Haraway's axiom, 'Make Kin Not Babies!' (Haraway 2016, 102), the movement away from anthropocentrism enjoins a redefinition of relationality, the subjects that constitute connection. For Haraway, 'kin' refers to 'something other/ more than entities tied by ancestry or genealogy ... making kin and making kind (as category, care, relatives without ties by birth, lateral relatives, lots of other echoes) stretch the imagination and can change the story' of human hegemony (Haraway 2016, 102–3). Thus, the critical review of having children is not an inducement to privation and austerity, but the exploration of other relational and communal possibilities.

Conclusion

Given the rapid decline of biodiversity, the destruction of ecosystems, and rampant global heating, educators can play a role in facilitating conversations around the ethics of having children as a part of responsible parenthood and citizenship. Although this proposal swims against the current of established norms, the educational effort to curb population growth remains preferable to government policies that seek to regulate human reproduction. Teachers can demonstrate pedagogical finesse in provoking thoughtful reflection without demolishing students' faith in human possibilities. At the very least, textbooks should present discussions of the ecological impacts of having children 2017, especially for families who maintain middle- and upper-class lifestyles in rich countries. An investigation of the impacts of having children, a culturally sanctioned act that often eludes examination, does not enforce a bleak view of humanity but rather calls to the capacity for conscientious action, an awareness that is apt to rise to what the world requires of those who have the most significant ecological impact, and that generates the potential to forge kinship beyond the boundaries that separate species. This educational endeavour is not fundamentally anti-human but calls us to a more considerate and compassionate human life.

Notes

1. There is a tendency in public and academic discourse to use 'climate change' and 'ecological crisis' interchangeably. Climate change is one (but by no means the only) aspect of the ecological crisis, though many forms of environmental damage are interrelated. In this paper, I shall use 'climate change' to denote the anthropogenic warming of the atmosphere from greenhouse gas emissions; I shall use 'ecological crisis' to refer to the larger set of ecological harm that results from human activity.
2. Online repositories of historical satellite photos, such as GIS Lounge (Dempsey 2013) and LearnArcGis.com ('Learn ArcGIS' n.d.), can help illustrate the changes in land-use over time.
3. Critics might contend that this activity is premised on scarcity and limited resources. Ecomodernists will argue that 'the degree to which there are fixed physical boundaries to human consumption, they are so theoretical as to be functionally irrelevant' ('An Ecomodernist Manifesto' n.d.). This is not the position that I hold, but the disputation of this point is beyond the scope of this paper

Acknowledgments

The author would like to thank all the reviewers and conference participants who provided suggestions and feedback for this paper.

Disclosure statement

No potential conflict of interest was reported by the author.

References

Berry, T. 2006. "Evening Thoughts: Reflecting on Earth as Sacred Community" In edited byM. E. Tucker, 1st: Sierra Club Books

Carrington, D. 2019. "Microplastics "Significantly Contaminating the Air", Scientists Warn". *The Guardian*, 14 August, sec. Environment. https://www.theguardian.com/environment/2019/aug/14/microplastics-found-at-profuse-levels-in-snow-from-arctic-to-alps-contamination

Chiras, D. D. 2014. *Environmental Science*. Jones & Bartlett Publishers.

Curry, P. 2011. *Ecological Ethics*. 2 ed. Cambridge, UK: Malden, MA:Polity.

Dasgupta, P. S., and P. Ehrlich. 2019. "Why We're in the Sixth Great Extinction and What It Means to Humanity." In *Biological Extinction*, edited by P. S. Dasgupta, P. Raven, and A. McIvor, 262–284. Cambridge: Cambridge University Press.

Dempsey, C. 2013. "Timelapse Satellite Imagery – View Changes on Earth over Time". *GIS Lounge* (blog). 9 May. https://www.gislounge.com/timelapse-satellite-imagery-view-changes-in-the-earth-over-time/

'An Ecomodernist Manifesto'. n.d. "An Ecomodernist Manifesto". Accessed 22 January 2020. http://www.ecomodernism.org

European Union. 2019. "Fossil CO2 and GHG Emissions of All World Countries : 2019 Report". Website. Publication Office of the European Union. Publications Office of the European Union. 26 September. http://op.europa.eu/en/publication-detail/-/publication/9d09ccd1-e0dd-11e9-9c4e-01aa75ed71a1/language-en

Ewing, B., D. Moore, S. Goldfinger, A. Oursler, A. Reed, and M. Wackernagel. 2010. *Ecological Footprint Atlas 2010*. Oakland: Global Footprint Network.

Haraway, D. J. 2016. *Staying with the Trouble: Making Kin in the Chthulucene*. Durham: Duke Univ Pr.

Kant, I. 1993. *Grounding for the Metaphysics of Morals: With on a Supposed Right to Lie because of Philanthropic Concerns*. 3rd ed. edited by J. W. Ellington. Indianapolis: Hackett Pub Co.

Lange, H., and L. Meir. 2009. *The New Middle Classes: Globalizing Lifestyles, Consumerism, and Environmental Concern*. New York: Springer.

Leah, H. 2018. "Driving into the Future: The Impact of Self-Driving Haul Trucks on Oil Sands Mining Giant, Suncor Energy". *Technology and Operations Management* (blog). 18 November. https://digital.hbs.edu/platform-rctom/submission/driving-into-the-future-the-impact-of-self-driving-haul-trucks-on-oil-sands-mining-giant-suncor-energy

'Learn ArcGIS'. n.d. Accessed 7 December 2020 https://learn.arcgis.com/en/projects/depict-land-use-change-with-time-enabled-apps/

Leopold, A. 1986. *Sand County Almanac*. Reprint. Ballantine Books.

Litfin, K. 2013. *Ecovillages: Lessons for Sustainable Community*. 1 edition. Polity. 'Living Planet Report'. 2020. World Wild Life Fund. 2020. https://livingplanet.panda.org/en-us/

Orr, D. W. 2004. *Earth in Mind: On Education, Environment, and the Human Prospect*. Washington: Island Press.

Reuters. 2019. "Scientists Shocked by Arctic Permafrost Thawing 70 Years Sooner than Predicted". The Guardian. 18 June 2019. http://www.theguardian.com/environment/2019/jun/18/arctic-permafrost-canada-science-climate-crisis

Sachs, J. D., S. A. Karim, L. Aknin, J. Allen, K. Brosbøl, G. C. Barron, P. Daszak, et al. 2020. "Lancet COVID-19 Commission Statement on the Occasion of the 75th Session of the UN General Assembly." *The Lancet* 396 (10257): 1102–1124. doi:10.1016/S0140-6736(20)31927-9.

Singh, M. 2019. "It's Raining Plastic: Microscopic Fibers Fall from the Sky in Rocky Mountains". The Guardian, 13 August 2019, sec. US news. https://www.theguardian.com/us-news/2019/aug/12/raining-plastic-colorado-usgs-microplastics

Smail, K. J. 1997. "Beyond Population Stabilization: The Case for Dramatically Reducing Global Human Numbers." *Politics and the Life Sciences* 2: 183–192.

Todd, S. 2000. "When Is Guilt More than Just a Petty Face? Moving from Liberal Guilt toward Reparation and Responsibility in Education." *Philosophy of Education 2000*. 357–364.

Waters, C. N., J. Zalasiewicz, A. D. Colin Summerhayes, C. P. Barnosky, A. Gałuszka, A. Cearreta, et al. 2016. "The Anthropocene Is Functionally and Stratigraphically Distinct from the Holocene." *Science* 351 (6269): aad2622. doi:10.1126/science.aad2622.

Watts, J. 2020. '"Sleeping Giant" Arctic Methane Deposits Starting to Release, Scientists Find | Climate Change | The Guardian'. https://www.theguardian.com/science/2020/oct/27/sleeping-giant-arctic-methane-deposits-starting-to-release-scientists-find

Wheary, J. 2010. "The Global Middle Class Is Here: Now What?" *World Policy Journal* 26 (4): 75–83.

Wynes, S., and K. A. Nicholas. 2017. "The Climate Mitigation Gap: Education and Government Recommendations Miss the Most Effective Individual Actions." *Environmental Research Letters* 12 (7): 074024. doi:10.1088/1748-9326/aa7541.

Got milk? From growing strong bones to nurturing idealized subjectivities

Samantha Deane and Annie Schultz

ABSTRACT

Philosophers of education have written about the moral, ethical, racial, and gendered dimensions of the hidden curriculum of what we eat, who we eat with, and the significance afforded this moment of the school day. To this body of literature, we add the observation that female bodies were positioned by Jean Jacques Rousseau as necessary food for the stuff of society. We trace the ways in which Rousseau's rendering of the *natural* female body have followed us into our modern educational systems such that any hopes for gender equality in our educational and social structures must attend to the pervasive *naturalization* of the bodies – human and nonhuman – that we are taught to milk for all that they are worth.

In 1762, Jean Jacques Rousseau began his educational treatise *Emile* as a letter addressed to a mother concerning the education of boys which proclaimed the mother's duty to breastfeed her children (Rousseau 1979, 45). According to Rousseau, a male child's education begins at his mother's breast. As the primary educational subject, Emile, the exemplar male pupil, is guided by nature to drink deeply, wander, explore, and experience, and in so doing develop a strong and righteous subjectivity. Meanwhile, his female counterpart Sophie (who we can assume is also fed at the breast of her mother, though no explicit mention of her infancy is articulated) is confined to learning within the home and through the social norms of domesticity until she meets her mate, Emile, who teaches her about transcendence. Ever skeptical about looming industrialization, Rousseau aims to bring education back to nature; what is natural is what is best and right. Sophie is naturally bound to care for the home, to breastfeed her babes, and to nurture society's future reproducers. Emile is destined for citizenship, freedom, and independence.

In this paper, we are interested in the social and political consequences of the naturalization of the care work of women and one important symbolic representation of this work: milk. We start with Rousseau's *Emile* because Rousseau's characterization of the necessary, yet largely unacknowledged, female body pervades western educational thought. It has been well established by philosophers of education that schools function due in large part to the labor of women; not to mention, the children who arrive to schools are also largely products of the care of women. In this essay, we draw a connection from the breastfeeding mother who nurtures the masculinized subject to conquer the world and the dairy cow whose milk stands in for the absent breastmilk. We argue the presence of the milk carton continues to position women in the roles of care work even when the woman is not present and long after a babe has been weaned. In fact, because the milk is neatly packaged on a lunch tray, the milk-drinking subject is able to forget about the labor of the mother and of the dairy cow: two beings who experience intersecting modes of oppression and subjugation.

As we see it, from the beginning, boys are educated to 'milk' the care of women for all its worth. They are taught that women exist to nurture their subjectivity at the expense of their own. When the child reaches school age, while the care work of women is ever-present, the breast-feeding aspect of care subsides, and the child increasingly receives their nutrition outside the home: typically, at school. One major component of the schools' attempt to nourish children in the lunch hour is packaged cows' milk. Milk in schools is not given much thought – it is absent from much of the literature in educational theory – but we argue that it holds symbolic power. In what ways does the trivialized labor and sacrifice of the dairy cow intersect with our gendered notions of the *natural* roles of women in care work and education? Both beings do a job assumed to be inherent in their nature. While human women exist in much more desirable circumstances than dairy cows, we argue their work might be similarly trivialized as it is chalked up to what is *natural*. Hence, in this paper we are giving attention to milk in school cafeterias as a gendered object with important iconographic consequences.

The milk carton is ubiquitous in school cafeterias. Prior to and now following the Obama era, milk often was, and now is again, the only beverage offered to children who receive low and reduced price lunch in schools (Robinson and Mulvany 2019). Within the 200 USD billion US dairy industry, 'schools are a major purchaser of milk, accounting for approximately 470 million gallons in the United States alone' (Weaver-Hightower 2011, 15–21). In fact, during the 2020 coronavirus pandemic, shuttered US schools resulted in euthanized cows, adjusted lactation cycles, and loads of dumped milk (Skerritt and Hirtzer, 2020). It would seem the affects go both ways; schooling impacts the life of the dairy cow just as the dairy cow's milk affects the body of the school child. Milk has also influenced popular culture and childhood play: the *Got Milk* campaign of the 1990s enticed everyone from celebrities like Kermit the Frog to supermodel Naomi Campbell to pose with milk mustaches. The same campaign spurred toy maker Mattel to create milk-related Barbie and Hot Wheels toys and author Caroline B. Cooney to pen a young adult book series featuring the milk carton. Cooney's series, *The Face on the Milk Carton*, was based on the National Child Safety Council's milk carton advertisements for missing kids, and it captured the attention of every 10 year old precisely as 'Got Milk' garnered the national attention of adults and children alike (Palmer 2012; Kardashian 2014; Algar 2019). In American society, milk, childhood, and education are undeniably connected, yet the relationship between milk and education is largely ignored.

What is milk doing in educational spaces? Many children have allergies to or difficulty metabolizing milk, so the argument that it is served in schools as a healthy alternative to soda or sugary juices is a fallible one. In fact, the science supporting milk's connection to human health and development has been challenged (Thorning et al. 2016). Moreover, beyond health concerns for individual children, ecofeminist scholars such as Greta Gaard have pointed out the systemic racial and colonial history of milk production and consumption globally (Gaard 2013; Sainath 2012; Kemmerer and Adams 2011; Harper 2009).

Ecofeminism is further concerned with the degree to which animals exploited for agricultural industrial purposes are feminized. For example, Carol Adams and Josephine Donovan (1995) have pointed out that species, as a locus of hierarchy, resembles gender and race, two areas which feminist theory devotes itself to. Clearly, there is something problematic in the comparison of the function of the dairy cow to the selfless acts of care given by women to children and communities; both because women-animal comparisons are dehumanizing and because of problematic hierarchies of sentiency, which reduce cows and other animals to an abject bestial status. Taking an intersectional approach to sexism and speciesism, as well as linking milk-making bodies to ideas about gender and nature, this essay brings together milk making, education, and a gendered political economy.

In the case of the care work of women, the role of mothers and the task of education was conflated from the start, such that it is difficult to disentangle them now (Boydston 1994; Coontz 1993, 65). Parsing our notions of what it means to become a subject from our gendered and naturalized assumptions about milk consumption is integral to an educational project that aims toward liberation. In the case of the dairy cow, our argument is not that children need to come face-to-face with the dairy cow, as many children already do, but, rather, that in America our ideas about education, citizenship, and schooling are inextricably connected to milk. Therefore, embarking upon the joint study of milk and education makes visible the care work of women, the gendered subjects of education, and the ways in which milk may work as a problematic symbol of oppression.

We aim to situate the educational facets of breastfeeding and the work of attending to children's bodies, which happens at home but prepares (male) children for life in public, within a gendered political economy. Leaving Rousseau and 18th century France, we then travel to 20th century American society to connect Rousseau's educative concern for breastfeeding to the labor of mothers and wives, which is often swiftly forgotten upon the child's entry to the public space of school. Gender is at the heart of Rousseau's thoughts on natural and naturalized education, though once we reach the 20th century it becomes possible to speak of the gendered political economy that is built into liberal democratic capitalism. Here we find women both having and nurturing children and yet joining men as democratic and economic subjects. In this second section of the paper, we discuss the ways that milk is positioned in schools and culture in ways that work to diminish the economic and social value of care work. In the third and final section of the paper, we explore the hidden curriculum of the milk that is consumed during the school lunch hour by engaging with ecofeminist theory on the woman-cow comparison: both beings are uncompensated for their labor and trivialized in our cultural iconography.

Rousseau and the education of male subjectivity

For Rousseau's Emile, 'living is the job' (Rousseau 1979, 41). Setting the stage for his male student's singular education in the job of life, Rousseau begins his treatise, *Emile*, with a series of instructions about the obligations of women. Addressing his female reader, he pleads:

It is to you that I address myself, tender and foresighted mother, who are capable of keeping the nascent shrub away from the highway and securing it from the impact of human opinions! Cultivate and water the young plant before it dies. Its fruits will one day be your delights. Form an enclosure around your child's soul at early date. Someone else can draw its circumference, but you alone must build the fence (Rousseau 1979, 38).

The duty of the mother, Rousseau tells us, is to feed and nurture her infant and to see that his potential to grow is not corrupted by society until an educator arrives. For her care and effort, she is rewarded when her child becomes a man for himself. Although Rousseau's primary concern and subject is Emile's development, he begins with the human who births Emile. The biological and material having of children requires a woman's body. Hence, Rousseau continues, 'There is no question about the duty of women. There is a dispute as to whether, given the contempt they have for it, it makes any difference for the children to be nursed with the mother's milk or that of another' (Rousseau 1979, 45). In other words, women have a biological and therefore natural

obligation to nurture infants with the milk from their own body – to cultivate and water the young plant. On this, as with everything else, we must follow the dictate of nature, insists Rousseau. On Rousseau's interpretation of nature's commands, women are born to bear children and their bodies are constructed to ensure the survival of their offspring. Hence, when mothers agree to nurse their children, 'morals will reform themselves, nature's sentiments will be awakened in every heart, the state will be repeopled' (Rousseau 1979, 46). Another woman's milk is not sufficient, neither is the milk from an animal. Rather by drinking from the breast of the mother, the son develops a formative and necessary attachment to his mother and the mother to her son. In fulfilling her obligation to nature, the mother creates a domestic life that sustains affection and guards her young son from the bad morals of men and things. On Rousseau's view, breastmilk is quite literally the food on which a moral nation emerges.

Emile is nurtured with breastmilk from his mother's domesticated body in infancy so that he may bound about nature until it is time for his subjectivity to bloom. Free to follow nature's laws, Emile is educated to become an individual who has absolute control over his psychological dispositions. He is raised away from the corrupting influences of other humans and his environment is carefully structured to keep unnecessary habits from setting in and subverting nature's aims. As a child, Emile is happy and free from the tyranny of opinion (Rousseau 1979, 162). When he is able to comprehend the future and the past, Emile's capacity to reason becomes evident and the tutor begins his slow introduction to the society of men. Emile's education shapes him into the perfect individual. 'He considers himself without regard to others and finds it good that others do not think of him. He demands nothing of anyone and believes he owes nothing to anyone' (Rousseau 1979, 208). Emile's education is dedicated to the production of self-love which alone guards against false comparisons to other men (Rousseau 1979, 235).

Corrupt bodies, or bodies that force us to attend to our limits, damn the education of the soul, says Rousseau (Rousseau 1979, 53). Thus, female bodies with their constant demands via a monthly menstrual cycle exclude women from the soul's education, while fit and well constituted male bodies that have been properly nurtured are the ideal educational subjects. Thus, Emile's female counterpart Sophie is molded through public opinion, convention, and familial observation to fulfill her duty to be a wife and mother. Sophie is educated to obey Emile, as, for Rousseau, the job of the woman is 'To please men, to be useful to them, to make herself loved and honored by them, to raise them when young, to care for them when grown ... to make their lives agreeable and sweet – these are the duties of women at all times, and they ought to be taught from childhood' (Rousseau 1979, 365). As Jane Roland Martin (1981) argues, Sophie is the key character here. It is her work as mother and wife that nurtures the subjectivity of men and that ensures the continuity of the state. Indeed, this is Rousseau's primary concern. Without women who are willing to raise and nurture their male children, the family dissolves, and without the family our fundamental moral affections are snuffed out (Rousseau 1979, 363).

Convince women to have children, at the appropriate intervals, to feed their babies from their own bodies, to obey their husbands, and a moral civilization has a chance. After all, Emile is not reared to be a lone woodsman. His individualism is the cornerstone of his capacity to be a democratic citizen.

Emile is educated for the political realm. Sophie is educated to sustain Emile. In Martin's words, 'That Sophie's place is not the political realm, but the home, is itself, then, a political commitment of Rousseau's' (Martin 1981, 371). Indeed, the feminist philosophical canon identifies this role as the earliest trapping of the woman in a state of imminence. Simone de Beauvoir identifies the role of motherhood specifically as one in which, only through its yield of

the transcendent subject, the child, can the woman reach some state of transcendence through the act of perpetuating life.

Our argument, like Martin's, is that Rousseau's sexist political commitments are still present in our educational thought and practice. Extending Martin's premise and drawing on Anne Crittenden, we agree the experiences of women have been ignored by educational theorists and that the male subject has become the generic universal; however, on our view, the domestication and exploitation of women's *bodies* lies at the core of our educational imagery.

We cannot attend to the experience of women nor the problematic universality of the male position, without the milk of mothers. To paraphrase Crittenden, like infants in cribs, the entire educational enterprise takes female caregiving utterly for granted.

Milking women and the ontological significance of milk

Rousseau begins with the mother and her breasts; he is right to do so. The caring labor of mothers is vital for human continuity. The physical and material work of the female body to birth and nurture a child is both natural *and* labor. For many women, the experience of having and rearing a child, including breastfeeding, is a pleasurable one and something looked forward to, chosen, and cherished. Yet, Rousseau's myopic focus on the naturalness of a mother's care has resulted in a society that unduly profits off the *labor* of mothers. Although many women choose to be mothers or care givers, to naturalize this choice is to trivialize it and diminish its social, political, and economic value. It is time we recognize the complex ways the bodies of women and the labor of mothers are being rendered invisible by the ever-present milk carton and its symbolic reminder that among humans only women are yoked to their bodies (though animals like the dairy cow are always and already bodies, never minds).

Our argument hinges on the ways in which care work, which has been largely done by women and subsequently feminized, has been coded as natural and thus removed from political consideration. If we trivialize the body and what is 'natural,' we by extension trivialize women and the work they do to make and sustain society – political and economic work. We argue that the milk students drink as a routine part of their lunch period is more than milk.

We can begin to acknowledge and problematize the symbolic power of milk by first taking it seriously as an object worthy of our attention. From the point of view of the students in a cafeteria, the milk is just there. However, it is not *just there*: it is the result of an ideological narrative that positions beings who menstruate as unsuited for subjectivity. And, further, the presence of milk in schools is the result of factory farming, commercialized industrial corporate distribution, unpackaging, preparation, supervision and service on the part of the cafeteria staff (statistically mostly women) (Weaver-Hightower 2011). Similarly, the arrival of the children themselves to the school building fed, nourished, grown, clothed, socialized to behave in a particular way, is also characteristically the result of the work of mothers and women. Therefore, the milk in the cafeteria and the care of mothers and women are ontologically similar in their material representations.

If we consider that the milk in the cafeteria might be an ontological representation of what is *natural* in the care work of women, we run into some problems. Though women's bodies have the biological propensity to bear children just as men's bodies have the biological capacity to produce sperm, the decision to have children and the work of care for children is just that: a decision about material life in society. Contrary to Rousseau, there is nothing 'natural' about female bodies that is not also true about the bodies of men or cows. Though far from 'being natural or traditional,' female bodies and bodily practices have been positioned as such: to serve, though not appear in, the perpetuation of a public body. Just as 'the use of chemical and petroleum-based fertilizers and pesticides' toward the mass production of cows' milk is not *natural*, we ought to question the extent to which the role of women to nurture and nourish the idealized subject which is symbolized by the milk on lunch trays is *natural*. Put another way, though we see cows' milk as a natural substance that provides nourishment to a growing body because it comes from another natural being (a cow), we know on some level that there is nothing natural about the way cows' milk is produced. Although yielding milk is an inherent biological capacity of the dairy cow, systematically breeding female cows, confining them to stalls, attaching them to automatic milking systems, and commercializing their yield as a consumable product is not an act of nature: it is an act of the industrialized political economy. Similarly, just as childbearing, breastfeeding, and rearing is an inherent biological capacity of a woman, does not mean that these acts should be discounted from the political economy on which our society depends because they are *natural*.

To be sure, not all women are involved in the reproduction of humans or the caring economy. As the journalist Ann Crittenden points out, 'women no longer have to look to a man to fulfill their dreams of accomplishment or economic security.' With modern medicine the same could be said about having a child. Women no longer need the physical presence of a man to bear and have a child. Yet, she continues, 'With the arrival of a child, a mother's definition of accomplishment becomes more complex, her workload goes up, and her income and independence go down' (Crittenden 2010, 26). In other words, though a woman can do it all, should she choose to bear children partnered or not, her life becomes infinitely more challenging. The data bears this out. Mothers are more likely to end up in poverty than childless women or men. Mothers work the longest hours and go to more teacher conferences, school meetings, and little league games than fathers (Crittenden 2010, 25–26). In spite of these odds or because of them, women still choose to have children, to become mothers, to love their children, and to care for them. This, however, does not mean their labor should be taken for granted or that women who seek renumeration for their contribution to a nation's human capital stores, for keeping babies alive, clean, and healthy, for raising kind citizens, and for attending all of the school meetings, should be cast as *unnatural* (Crittenden 2010, 64). To be clear, we are not suggesting that the work of care or motherhood are not worthwhile ways to spend one's life and time.

Rather, we are suggesting that this work is rendered trivial and invisible because social, cultural, and educational institutions view it as natural.

There was another path. In the early 20th century, a debate between Charlotte Perkins Gilman and Anna Howard Shaw set the tone of the women's movement for the rest of the century (Crittenden 2010). According to Gilman, within the economic logic of capitalism the wife/mother is equivalent to a domestic servant who can only gain independence with the abolishment of the nuclear family (Gilman 2018). Drawing on a complex rendering of species development (and the pseudoscience of eugenics) Gilman argues that women have become the subject of an excessive sex distinction and to make her point she too turns to the cow: 'To make clear by an instance the difference between normal and abnormal sex-distinction, look at the relative conditions of a wild cow and a "milch cow," such as we have made. The wild cow is female. She has healthy calves, and milk enough for them; that is all the femininity she needs . . . We for economic uses, have artificially developed the cow's capacity for producing milk. She has become a walking milk machine . . . the secretion of milk is a maternal function – a sex function' (Gilman 2018, 44). The lesson is that women ought to be treated like the wild cow not the milch cow. For Gilman, the choice is between natural/wild and domesticated/sexed. What is key about Gilman's proposition is her denigration of childrearing under capitalism (in her speculative utopian novel, *Herland*, motherhood constitutes a society of only women who reproduce through asexual means). In her words, 'there is no equality in classes between those who do their share of the work in the largest, newest, highest ways and those who do theirs in the smallest, oldest, lowest ways' (Gilman 2018). The lowest work equates to the labor in service of the family. Thus, Gilman's radical proposal to abolish the family was based on a view that the work to sustain children and make homes was not valuable work: it was old work – work that had sustained communities for centuries but work that did not advance society. The gambit, that women learn to reproduce via asexual means in order to become free range mothers, or, to put it another way, that women become like men as full citizens and members of the commercial work force, set us on the wrong path.

Gilman lost the debate but won the century: the devaluation of motherhood as a normative product of capitalism saw women enter the 'public' workforce to shed the label 'economic dependent.' Ann Howard Shaw, on the other hand, argued that wives and mothers do important work in creating a home and raising children for which they should be compensated. Known for her oratory and activism within the National Women's Suffrage Association (NWSA), scant attention has been devoted to her ideas. In contrast to Gilman and many of her suffragette contemporaries, Shaw was a poor immigrant who lived outside of the heteropatriarchy with her partner Lucy E. Anthony. In the scholarly biography, Trisha Franzen convincingly argues that Shaw's experience with poverty and varied experience of gender, influenced her leadership of the NWSA and advocacy (Franzen 2014, 80). In her 1908 debate with Gilman, Shaw argued that men and women are economically interdependent on one another and that governments ought to support, financially, the work of mothers. Less radical than her interlocutor, Shaw sought a world in which motherhood and care work are valued and supported for the material, economic, and political contribution to society. For Shaw, nothing about the labor of women is natural.

The naturalization of mothers' labor has granted American society a hall-pass to use, cultivate, and consume the fruits of women's bodies. On our view, the labor of mothers is political and public work and as such it is neither natural or unnatural – rather it is a decision a woman makes as a citizen and participant in the economy. Yet, because our educational institutions (and thus one might argue our society itself) are built upon a tradition which cannot account for a female body that bleeds and thinks, in Rousseau's lexicon, the corrupted body, we continue to teach children to ride freely on the bodies of women and *naturalized others*.

The school lunch

Although these *naturalized others* regularly appear in schools, we try at every turn to obscure their role. The trivialization of the body and its processes is a key part of the mind-body dualism that has pervaded Western ideologies since the last half of the eighteenth century.

According to this dualism, matters of the body are only included in the school day because students cannot think when the body's needs are ignored. The educational project continues to prioritize and glorify, the mind. Educational institutions, which privilege the mind as the disembodied brain in a jar that is uniquely apt for transcendence, too often ignore the educative importance of key moments in the school day when children are free to be in their bodies *and* engage, however passively, with the bodies of nonhumans and their byproducts; namely, *milk*. It is no wonder then, from the view of the institution, that what children consume in schools as a part of a grudgingly necessary lunch hour and the labor it might be a stand in for, both human and nonhuman, is not paid much mind (Weaver-Hightower 2011, 15–21).

Yet, educational theorists and journalists alike have puzzled over the significance of school lunch. For those of us looking at education, that lunch hour is a microcosm. Susan Levine, a historian, documents the civics training that emerges in one of the longest lasting public welfare programs, the school lunch. Philosophers of education Suzanne Rice and A.G. Rud argue that for children the lunch hour is a pivotal point in the day where social class becomes apparent, and, most importantly, where learning with and through the body happens (Rice and Rud 2018). Fellow philosophers of education Bradley Row and Samuel Rocha call our attention to the 'hidden curriculum' in the school day: 'food is routinely viewed as a mere accident to the so- called real learning that takes place in the classroom' (Rowe and Rocha 2015, 484). For Rowe and Rocha, the implications of the things consumed in the cafeteria 'extend well beyond and before the cafeteria itself' (Rowe and Rocha 2015). While Rowe and Rocha call for 'a serious intentionality toward' the food served in schools at lunchtime, we would like to propose a consideration of the bodies at stake in the arrival of milk on those lunch trays and the students in the cafeteria. To be sure, although educational research 'often focuses on women – who constitute the majority of teachers and principals – many aspects of women's labor in schools are hidden' (Weaver-Hightower 2011, 16). Echoing our colleagues, we argue that the lunch hour is a microcosm that illuminates our deeply held social, political, economic, and gendered assumptions about that which is natural, consumable, useable and that which is capable of transcendence. In fact, one omnipresent lunchroom food – the milk on student's lunch trays – works as a stand in for the embodied labor of women that schools and communities are often built on top of.

How do we get from breast milk to packaged cows' milk on a student's lunch tray? In this essay we have hinted at one path: a recognition of that which is constructed as natural as a part of a desire for control. In other words, because the dairy cow is seen as female, her labor to produce milk is registered as natural. The use of an animal comparison, to make sense of the trivialization of women as we suggest here with the dairy cow, is subject to question. As ecofeminists Carol Adams and Josephine Donovan explain, 'historically, the ideological justification for women's alleged inferiority has been made by appropriating them to animals,' and, moreover, that 'women's bodies have been seen to intrude upon their rationality' (Adams and Donovan 1995, 1). One notable set of pejoratives of women-animal comparison is to that of farm animals: Donovan and Adams point out that it is typically 'farm animals [that] are feminized ... in our cultural iconography,' and, specifically, that the 'exploitation of the cow for her milk has created a gender-specific image' (Adams and Donovan 1995, 6;13). In addition to the physiological comparisons made between the breastfeeding woman and the dairy cow, social pejoratives further equate the human woman's state of motherhood to the unfortunate state of the dairy cow: 'confined to a stall ... milking becomes something done *to* her rather than *by* her – she is seen as passive and dull' (Adams and Donovan 1995, 13). Popular media and social rhetoric equate the nursing mother with being confined to the space of the home (or when in public to the restroom or some enclosed lactating space) and relegated to a body whose sole purpose is to feed and sustain the life of her offspring. More broadly, the woman and the dairy cow share a position in the 'structures of industrial ... production' in which both do a 'job' they are uncompensated for, and the results of their labor are taken for granted (Gaard 2013, 598).

The unaccounted labor of maternity and breastfeeding is trivialized in American society to a similar degree as the suffering of dairy cows. In fact, we argue the reason that farm animals are feminized is *because* they are similarly trivialized. Although the effort of bringing the conversations happening in the fields of animal ethics and ecojustice to philosophy of education is a pertinent one, our primary aim has been to illustrate that the trivialized and uncompensated labor of two beings (women and cows) are parallel in that they are easily made invisible by the unquestioned consumption of milk in schools.

Gendered notions of subjectivity are emerging in unsanctioned educative moments – such as lunch. The milk on the lunch tray emphasizes and reinscribes the role female and nonhuman bodies *necessarily* play in the perpetuation of this unjust society. If milk were to disappear from lunch trays, might the ontological position it signifies dissipate in equal measure?

Metal cows and a conclusion

We have aimed in this paper to problematize the extent to which the care work of women should be taken for granted for the reason that it is 'natural.' The milk served in schools is symbolic of this naturalization in that it is also a material we have been culturally removed from, and, therefore, the story behind its abundance is rendered indivisible. Milk may seem the most 'natural' possible substance one might drink; after all, it comes from *nature*. However, milk 'is a historically unprecedented product of industrialization, urbanization, culture, and economics' (Gaard 2013, 597). Milk only reaches lunch trays after vast amounts of atrocity, both environmental and moral. Yet, we continue to use the image of the content mother cow, happy to remain in martyrdom for the nourishment of the other. The truth is that the milk in schools could not be further from the natural process that yielded it.

Milk has been both naturalized and gendered. Just as the inherent moral ineptitudes associated with serving milk in schools (as noted in the beginning of the paper: colonialist, labor exploitation, racialized health issues) is made invisible through routine and cultural expectation. As Greta Gaard points out, the cow and her milk are taken for granted in mass industrialized factory farming precisely because the cow is gendered: 'For too long, the dominant culture has childishly projected its own gendered image onto nature as selfless and self-sacrificing mother' (Gaard 2013, 613). Attending to the milk in our schools challenges us to recall the construction of what is natural around that which humans hope to exploit. Female bodies, with their all-too- physical states, were positioned as natural and corrupt; powerful and weak; but, most of all, as necessary food for the stuff of society.

When we, humans, understand something as natural it becomes quite difficult to make sense of alternative visions of ontological ordering. Thus, in this essay, two opposing narratives surrounding women's bodies and reproductive capacities are at play. One is of domestication and subjugation. The other is of life-giving empowerment. Though these narratives appear distinct, they are intimately connected in the construction of *naturalized bodies*. According to Rousseau, it is a woman's life-giving power that chains her to her body. Sophie's subjugation to public opinion is demanded by her biological ability to give birth and duty to ensure paternity. *Nature* for Rousseau turns on a vision of democratic society wherein each gender has a role to play in the making of citizens. Ecofeminists have set out to challenge biological determinism by analyzing the discursive practices that render bodies useful and intelligible. A choral refutation of Rousseau is audible if one knows how to tune in. This, however, is a matter of educational significance that teachers, policymakers, and parents must hear. If our educational visions intend to bring about more equitable, just, humane, and egalitarian relations, then we must stop 'milking' the care of women and female bodies and start thinking about the ontological

significance of the products children consume. After all, milk does a body good, but only at the expense of other bodies.

Disclosure statement

No potential conflict of interest was reported by the authors.

References

Adams, C. J., and J. Donovan, edited by. 1995. *Animals and Women: Feminist Theoretical Explorations*. 1st ed. Durham: Duke University Press.

Algar, S. 2019. "New York City Schools Want to Ban Chocolate Milk." *New York Post*, 15 September 2019. https://nypost.com/2019/09/15/new-york-city-schools-want-to-ban-chocolate-milk/

Boydston, J. 1994. *Home and Work: Housework, Wages, and the Ideology of Labor in the Early Republic*. 1st ed. 30-56, New York: Oxford University Press.

Coontz, S. 1993. *The Way We Never Were: American Families and the Nostalgia Trap*. Reprint edition. New York, NY: Basic Books.

Crittenden, A. 2010. *The Price of Motherhood: Why the Most Important Job in the World Is Still the Least Valued*. In *10th Anniversary Edition*. New York: Picador.

Denworth, L. 2020. "The Outsize Influence of Your Middle-School Friends." The Atlantic. 28 January 2020. https://www.theatlantic.com/family/archive/2020/01/friendship-crucial-adolescent-brain/605638/

Franzen, T. 2014. *Anna Howard Shaw: The Work of Woman Suffrage*. Baltimore, UNITED STATES: University of Illinois Press. http://ebookcentral.proquest.com/lib/luc/detail.action?docID=3414330

Gaard, G. 2013. "Toward a Feminist Postcolonial Milk Studies." *American Quarterly* 65 (3): 595–618. doi:10.1353/aq.2013.0040.

Gilman, C. P. 2018. "Women and Economics A Study of the Economic Relation between Men and Women as A Factor in Social Evolution." http://www.gutenberg.org/ebooks/57913

Harper, A. B., edited by. 2009. *Sistah Vegan: Black Female Vegans Speak on Food, Identity, Health, and Society*. New York: Lantern Books.

Kardashian, K. 2014. "The End of Got Milk?." 28 February 2014. https://www.newyorker.com/business/currency/the-end-of-got-milk

Kemmerer, L. A., and C. J. Adams. 2011. *Sister Species: Women, Animals and Social Justice*. 1st edition ed. Urbana, Chicago, and Springfield: University of Illinois Press.

Martin, J. R. 1981. "Sophie and Emile: A Case Study of Sex Bias in the History of Educational Thought." *Harvard Educational Review* 51 (3): 357–372. doi:10.17763/haer.51.3.p814021448113257.

Palmer, B. 2012. "Why Did Missing Children Appear on Milk Cartons?." Slate Magazine. 20 April 2012. https://slate.com/news-and-politics/2012/04/etan-patz-case-why-did-dairies-put-missing-children-on-their-milk-cartons.html

Rice, S., and A. G. Rud, edited by. 2018. *Educational Dimensions of School Lunch: Critical Perspectives*. Palgrave Macmillan. https://doi.org/10.1007/978-3-319-72517-8

Robinson, P., and L. Mulvany. 2019. "Big Dairy Is about to Flood America's School Lunches with Milk." *Bloomberg.Com*, 9 January 2019. https://www.bloomberg.com/news/features/2019-01-09/big-dairy-is-about-to-flood-america-s-school-lunches-with-milk

Rousseau, J.-J. 1979. *Emile, or on Education. Translated by Allan Bloom*. New York: Basic Books.

Rowe, B., and S. Rocha. 2015. "School Lunch Is Not a Meal: Posthuman Eating as Folk Phenomenology." *Educational Studies: Journal of the American Educational Studies Association* 51 (6): 482–496. doi:10.1080/00131946.2015.1098643.

Sainath, P. 2012. "Cattle Class: Native Vs Exotic." *The Hindu*, 6 January 2012, sec. Sainath. https://www.thehindu.com/opinion/columns/sainath/Cattle-class-native-vs-exotic/article13355009.ece

Skerritt, J. and M. Hirtzer. 2020. "More Milk Cows Are Ending Up as Hamburgers in Latest Virus Twist - Bloomberg." Bloomberg.Com, May 7, 2020. https://www.bloomberg.com/news/articles/2020-05-07/more-milk-cows-are-ending-up-as-hamburgers-in-latest-virus-twist

Thorning, T. K., A. Raben, T. Tholstrup, S. S. Soedamah-Muthu, I. Givens, and A. Astrup. 2016. "Milk and Dairy Products: Good or Bad for Human Health? an Assessment of the Totality of Scientific Evidence." *Food & Nutrition Research* 60 (January): 32527. 1-N.PAG. doi:10.3402/fnr.v60.32527.

Weaver-Hightower, M. B. 2011. "Why Education Researchers Should Take School Food Seriously." *Educational Researcher* 40 (1): 15–21. doi:10.3102/0013189X10397043.

Psychoanalytic ecofeminist Dorothy Dinnerstein: Theorizing the roots of rapacity

Gregory Bynum

ABSTRACT
This article proposes that Dorothy Dinnerstein's philosophy can help us understand the problem of miseducation that places male-dominated and 'masculine' rapacity at the center of so many human endeavors, including capitalist economic exploitation and environmental exploitation. Dinnerstein argues that early childhood experiences of female domination lead to reactive and immature adult preferences for excessive, triumphing, rapacious, male rule. In Dinnerstein's theory, the solution to this psychologically deep-rooted problem is for men to do half of the childcare work. This article acknowledges and refutes arguments against Dinnerstein's theory, and expounds its soundness and benefits. Dinnerstein's thought can help us put a brake on the capitalist rapacity that threatens our very survival while increasing possibilities for human self-actualization.

Introduction

The human-caused devastation of the world's natural resources presents a picture of rapacity and competition for monetary gain without adequate concern for fundamental values of sustainability, humaneness, and respect for human rights. In this situation what must be understood, challenged, and overcome is the rampant primacy of the non-values-based aim of winning contests for monetary profits in global business sectors including energy, agribusiness, and manufacturing. To this end, this article will pose and respond to the following question: What are the psychological roots of humanity's dangerously rapacious, non-human-values-based approach to the natural environment?

It will be proposed that a promising approach to theorizing and developing our response to these questions is provided by Dorothy Dinnerstein's ecofeminist philosophy as articulated in her classic 1976 work, *The Mermaid and the Minotaur: Sexual Arrangements and Human Malaise*. Dinnerstein's text can help us understand the problem of mis-education that places male-dominated and 'masculine' rapacity at the center of so many human endeavors. Her proposed explanation of the source of unbridled, collectively undertaken, dangerous and excessive human rapacity is as follows. In early childhood, peoples' understandings and feelings about the justified uses of human power become dangerously mistaken as a result of a series of psychological events beginning with women's domination of early childcare. Since, for most people through most of history, the relationship with another human being that comes closest to feeling like a relationship with a deity is the relationship between a small child and one's female caregiver, our lifelong quests to become fully human beings in fully human relationships with one another are excessively determined as reactionary against women's power as experienced during our earliest years. These quests tend to take the form of emotionally and psychologically understandable, but disastrously misguided, efforts to artificially and vindictively diminish the power of women and the significance of aspects of experience associated with women and deemed 'feminine' (caring, nurturing, physicality unintegrated with intellectual reflection, responsible responsiveness to vulnerability in human beings and in other living beings), while also – in a reactive move in which we misguidedly feel that extreme triumphs of male and 'masculine' power entail, for us, humanizing, maturity-achieving, and mastery-achieving movements out of the resented power of female adults as experienced in childhood – artificially and vindictively over-exalting male power and masculine power. Since men's ordinary, worthwhile, human-needs-respecting achievements lack the deity-like qualities immaturely desired by individuals caught in the throes of their wishes for triumph over the seemingly deity- like, and intolerably infantilizing, power of women felt to be a continuation of infantile and early childhood experience, people's preferred enactments of adult power often are not enactments that display sensitive and rational attention to human needs well-observed and well-understood. Instead, people prefer triumphant and excessive enactments of masculine power – enactments that seem to perform god-like achievements of ever-expanding dominion and might.[1] Hence, human beings' oft-demonstrated tendency to conform with, and energetically and enthusiastically to participate in, projects of massively heedless, dangerous, and deadly rapacity such as the devastation of the earth's natural environment.

For Dinnerstein, an indispensable step towards rectifying this situation is men's taking over half of the childcare beginning in the period of earliest human development. She proposes that, while we currently find it easy to immaturely scapegoat women and over-exalt men as a way to express our lingering feelings about being so completely dominated by another human being in early childhood, such self-expression would become psychologically unsatisfying if, for the average human being half of the time, a man and not a woman were the one to dominate us when we were at our tiniest, most vulnerable, most inexperienced, and most dependent. Since we, collectively, would no longer find it interesting to live lives motivationally rooted in immature reaction against the power of women (because men and women equally would invoke our reactionary fears of being re-infantilized by another human's power), we would find it necessary to see our feelings of fear of infantilization for what they are – irrational and immature fears that must be outgrown in order to for us to live sensitive, sensible, rational, non-rapacious, human (not godlike or god-following) lives, individually and collectively – lives that are responsibly responsive to human vulnerability and to the vulnerability of non-human living beings with which humans are vitally interconnected. Such an outcome, Dinnerstein indicates, is what is needed in order to slow, stop, and reverse our current trend towards human beings' patriarchy-driven decimation of the global, natural environment and, by extension, of ourselves.

The remainder of this paper will be divided into two sections: a section presenting, and refuting, objections that may be made to Dinnerstein's theory; and a concluding section discussing her theory's implications for thought and practice.

Refutations of objections to Dinnerstein's theory

This section of this paper will present and refute possible objections to Dinnerstein's theory, and to Dinnerstein's psychoanalytic, ecofeminist critical approach.

One objection that has been made is that Dinnerstein's imperative to share childcare equally between male and female adults is heterosexist, or biased against homosexual, child-rearing couples, in that it calls for a society of households that each have one man and one woman, in which the man and the woman equally share childcare. This objection takes a needlessly narrow view of Dinnerstein, whose proposal actually does not specify compulsory heterosexuality or any prescription to be uniformly applied to all households, but rather indicates a need for a society where *on average* children would be raised equally by male and female childcare providers. In this society there would be no reason for excluding families with two male childcare providers, families with two female childcare providers, or families in which male-female couples allocate more childcare either to the male or to the female partner. Rather, the social transformation Dinnerstein indicates is one to a society where *in general and on the whole*, in practice and in the expectations of childcare work that are placed upon individuals, there is no bias towards having any one gender do the majority of childcare. All individuals would equally be educated and encouraged in preparation for childcare work.

Another objection is that Dinnerstein presents a picture of human society in which everyone is male or female, which does not take into account the existence of intersex people (individuals born with bodies that are neither male nor female) or the existence of transgender people who identify as gender non-binary. In response, it should be noted that Dinnerstein at no point insists that all human beings are either male or female. Further, the pressures to conform to stereotypical 'masculine' and 'feminine' roles that Dinnerstein opposes are pressures that harm not only male and female people, but also (and pronouncedly) intersex people and gender non-binary people. Indeed, there is a history of intersex people being pressured into, and non-consensually subjected to, medical operations to 'change' their bodies into female bodies even though they are not female people. And gender non-binary, transgender people face psychologically harmful, social pressure to conform to expectations that they should be 'male' or 'female.'

An objection also may be made that men may not have the same innate, nurturing capacity as women. And some people feel that women's experience of carrying a child in the womb gives them a special connection with children that men lack. There also has been a concern raised that men could be too non-protective in their imposed role as nurturers, and some anecdotal evidence exists relating to this idea (Magary 2013; Hodgkinson 2010). Refuting these concerns, however, studies of parenting indicate that one's gender does not make one innately superior or inferior as a nurturer (Fine, 2010; Rivers and Barnett 2011). Further, some research indicates that socialization around communication about vulnerability is more supportive of the development of this ability in women than in men; and since responsible responsiveness to human vulnerability is an essential quality of good childcare, it may be that due to socialization, and not due to any innate gendered difference, men sometimes show a lack of skills needed for nurturing children (a lack that could surely be remedied by efforts being made to develop their skills) (Rivers and Barnett 2011).

From a historical perspective, it may be objected that, contrary to Dinnerstein's claim that excessive male power is dangerously unchecked, history is full of salutary objections to, and checks placed upon, excesses of male power, as noted in endnote 1 concerning the many stories about the overthrow of tyrannical kings who are supplanted by superior political leaders. In response to this, it should be noted that the critique and overcoming of these bad kings, while in sympathy with Dinnerstein's critique and in a way serving, perhaps, as part of a tradition in which it participates, are insufficient to overthrow the still-dangerous, still-undesirable, widespread preference for male authority and 'masculine' rapacity of the kind that Dinnerstein criticizes. And at worst, these tales of overthrown tyrants may serve to put at rest concerns that actually should remain lively and critically engaged in processes of social change and movement towards gender equality; excessive patriarchal power involved in pointless competition should be eliminated in all the spheres where it appears (family life, economic life, political life, managerial practice), and not just in the case of some historical, mythical male tyrants and other 'bad' male rulers whose existence serves, by contrast, all the more strongly to emphasize to legitimacy of those patriarchal, male leaders who are considered 'good' despite the fact that they themselves commit rapacious, authoritarian excesses.

As an objection to Dinnerstein from the perspective of the history of philosophy, it may be pointed out that the resolution of the male-female binary has been beautifully achieved for centuries beginning with ancient, Chinese philosophy that elegantly reconciles the 'yin' (feminine) and 'yang' (masculine) principles; and that Dinnerstein's writings put forward a problem that has already been addressed by centuries-old philosophical tradition. In response to this objection, it should be noted that Dinnerstein actually departs from this thought tradition by indicating a need for us to abandon altogether (not reconcile) traditionally established categories of 'masculine' and 'feminine' where the latter is passive, weak, soft, etc. while the former is active, strong, and hard, etc. What is needed is a change to a society where everyone is educated to feel, and to believe, that they can equally possess all of those qualities (strong, weak, active, passive, soft, hard ...) without gendered constraints limiting their sense of the possibility of experiencing these aspects of humanity in themselves and in others.

From a psychological and feminist perspective, it may be objected that Dinnerstein's choice to take a psychoanalytic approach is undesirable since Freud expressed some sexist views, and since his theory and other psychoanalytic theory over-emphasize early experiences of sexuality and sexuality-related trauma in ways that have been rightly rejected as invalid and unhelpful. In response to this objection, it may be stated that while one must certainly reject Freud's antiquated sexism and his widely contradicted way of conceptualizing the influence of sexuality in childhood, Dinnerstein actually does not make either of these Freudian errors.

Rejecting these negative aspects of the psychoanalytic tradition need not, and should not, entail entirely rejecting the following, fundamental and essential human question of psychoanalysis, which is the question that essentially concerns Dinnerstein: How does our childhood experience of power in human relationships affect our desires and decisions when we establish power relationships in adulthood in our families, in our institutions, and in our society? And one does not have to possess any loyalty to, or in-depth conversance with, any branches of psychoanalytic thought to seriously reflect and act upon Dinnerstein's driving, psychoanalytic question, which is: what are the likely implications for adult life of the fact that for most people, most of the time, early childcare is female-dominated?

Implications of Dinnerstein's theory

This portion of the paper will touch on several points of interest in Dinnerstein's thought – aspects of her thinking that, in distinctive and profound ways, offer resources for radically understanding, criticizing, and dismantling existing, patriarchal, rapacity-promoting and environmental-exploitation-engendering gendered social roles. The first point of interest is the way Dinnerstein centers women's power instead of men's power in her description of how male–female, masculine–feminine relationships develop educationally and socially. The second is the ways Dinnerstein can help us understand the gendered mis-education of people into traditionally 'masculine' and 'feminine' roles. The third point of interest is Dinnerstein's discussion of the traditional, so-called 'battle of the sexes' – a discussion that shows how the battle of the sexes is disempowering, destabilizing, and conservative, and a discussion that points to alternative approaches to promoting gender equality in education. And the fourth point of interest is that the undervaluing of women's intellect as part of resentfully diminishing women's power may be linked to undervaluing the intellects of the children who traditionally are under women's care and within 'woman's sphere.'

One benefit for education from Dinnerstein's thought is its potential for developing more deeply rooted and stable gender equality. More robust approaches are needed to challenge still-existing biases that favor men's and boys' intellectual and professional development and self-actualization over that of girls and women. For this purpose Dinnerstein's philosophy, with its focus on child-adult power relations, covers important ground that feminist thought sometimes neglects. Many feminists (importantly) focus on women achieving equal status in traditionally 'masculine' spheres of power. In contrast, Dinnerstein focuses on developing an imperative for men to take over half of the 'feminine' work of childcare. (We can think of the old saying that women will never have equality outside the home until men have equality inside the home.)

Interestingly, Dinnerstein's focus on decreasing women's power in their traditional sphere of childcare ultimately poses a profounder challenge to patriarchy than does a feminist approach that focuses only on women assuming more traditionally 'masculine' work. Unlike in patriarchal conceptions such as the Christian creation story in the bible's Book of Genesis, in Dinnerstein's conception, it is not male power but female power that is the source, origin, and prototype of all other power. For Dinnerstein *female* power exerted in childhood sets in motion the dynamics of patriarchy, and it is a change in *female* power – in the traditionally feminine work of childcare – that holds the greatest promise for bringing about the most profound and lasting change to gender equality in all spheres of human life. In this context, it is important to remember that children do experience their relationship with their first caregivers *intellectually* from a very early age; contrary to the expert opinion of a half-century ago, children under two years old are, in fact, capable of reasoning and understanding cause-effect relationships (Gopnik 2009, Hickling and Wellman 2001, Kushnir and Gopnik 2005, Kwak et al. 2015; Lucas et al. 2014).

I will next discuss my second point of interest, which is how Dinnerstein's philosophy can help us understand more deeply the gendered mis-education of boys, of men, and of people whom society seeks to categorize as 'masculine' (including women occupying 'masculine' jobs and other social roles) as well as the gendered mis-education of girls, of women, and of people whom society seeks to categorize as 'feminine' (including men occupying 'feminine' jobs and other social roles). This discussion is relevant for understanding how gender-biased expectations harm individuals who do not conform to conventional expectations of gender and sexuality including transgender people, lesbian and gay people, and intersex people. With this in mind, I request that it be taken as understood that when I describe processes that mis-educate boys, girls, men, and women, I am also implicitly criticizing gender-biased harms to gender non-binary people, to transgender people, and to intersex people.

As has been touched upon already, in Dinnerstein's view people are mis-educated into masculinity by the following process. When male power is privileged because of people's childish desire for it to triumph over female power which is resented as infantilizing, this immature privileging of male power does not carry with it any incentives for boys and men to achieve and demonstrate excellence in desirable qualities such as competence, compassion, integrity, knowledge, strong understanding, insight, or moral stature. All that matters in emotionally driven and immature patriarchal thinking is that boys and men be winners, regardless of the human destruction and the environmental destruction that men's victories may involve. As a result, we often see boys and men being miseducated to focus on defeating others, or on appearing to overmaster others, in various intellectual, managerial, economic, political, and other contexts, instead of being educated to achieve true excellence either in uniquely personal, individual self-actualization or in their ability to contribute to the well-being of others.

For Dinnerstein, there is also a process of mis-education to femininity – the mis-education of women and girls to fulfill the requirements of what she calls the 'shaky she-goddess' myth (Dinnerstein 1999, 208). Just as with mis-education to masculinity, there is an immature expectation of perfection when we mis-educate people to femininity. Girls and women are childishly expected provide the kind of perfect pleasure and perfect happiness that we remember our first childcare providers giving us when we were hungry and they fed us, or when they rescued us from what seemed to our inexperienced minds like impossible and eternal distress. Investment in this maternal she-goddess myth impedes personal development for everyone involved. It overburdens women with expectations that they provide perfect comfort and pleasure at the expense of attention to their own needs.[2] And it may hold back the people cared for in this way from becoming more emotionally mature and self-reliant, and from developing a capacity for non-codependent, adult relationships.[3]

I will now move to my third point of interest – Dinnerstein's discussion of the age-old, so-called 'battle of the sexes.' The title of her chapter, 'Sometimes you wonder if they're human,' is a phrase that is sometimes applied to men by women, and sometimes applied to women by men, in a process of mutual dehumanization (Dinnerstein 1999, 91). This mutual dehumanization, as Dinnerstein describes it, is the inevitable outcome of people's childish but deep-rooted efforts to divide humanity into male winners and female losers (touched upon above). The extreme separation and conflict between those considered 'masculine' or male, and those considered 'feminine' or female," is – to borrow language from John Dewey's *Democracy and Education* – possibly one of the most intractable barriers to free and full communication of experience between social groups that impedes fully democratic social relationships and fully democratic education (Dewey 1997, 99). A useful way to understand the 'battle of the sexes' from Dinnerstein's perspective is to consider the plot of 'The Taming of the Shrew' – a story of conflict between a man and a woman that was told in a famous play by William Shakespeare, and also in less famous versions of the story in which the defeat of the woman, Katherine, is much more physically brutal than in Shakespeare's version (Barton 1974, 106). Traditionally minded audiences, a Dinnerstein-inspired reading would argue, are immaturely gratified by this story because of the way it inevitably ends in the man defeating the woman after some spicy resistance on the woman's part; this gratifies people's lingering, anti-female resentment left over from childhood. Another part of the story's appeal may be its representation of the extreme alienation and conflict that arises in our society because men and women, who are intellectually equal, are childishly divided into the roles of masculine, dominant winners and feminine, submissive losers because of people's failure to outgrow their early, childish resentment of women's power. In our society today, the battle of the sexes is showing up in new, dehumanizing terms like womanplain, mansplain, momsplain, and toxic masculinity – words that invalidate people's thoughts on the basic of their innate natures as female and male human beings (and incidentally words that, when used in workplaces in the United States, arguably violate of Title IX in U.S. law because they single out individuals for discriminatory treatment on the basis of gender). A critique of these words informed by Dinnerstein points to the fact that, in the context of a society where people immaturely prefer to see men defeat women, a contest between the idea that men are inferior (mansplainers with toxic masculinity), and the idea that women are inferior (momsplainers and womanplainers), is rigged in favor of the idea of female inferiority, like the plot of *The Taming of the Shrew*. A better anti-sexist approach would ground itself in gender equality and in the increasing evidence that no-one's gender makes them innately superior in intellect, in nurturing capacity, in leadership ability, or in emotional intelligence (Rivers and Barnett 2011; Saini 2018; Fine 2010).[4]

This critique implies critical rejection of all examples of individuals creating and acting upon binaries in which male vs. female has connotations of good vs. bad, or bad vs. good. This implies criticism of some environmentalist thinking that identifies the earth and the natural environment as 'feminine' and 'good' while identifying human activity (despite the fact that humans are part of nature) as 'masculine' and 'bad.' A better approach, following Dinnerstein, would be to ground environmentalist imperatives in non-binary, non-gendered human needs such as the need for fairness and hope for the younger generation; the need for meaningful and respectful interaction with the non-human, natural environment; and the needs for order, integrity, efficacy, and moderation through the sustainable management of natural resources and human appetites.

Next, to move to my fourth point about Dinnerstein's thought, demeaning women out of childish resentment may negatively affect not only women, but also the children whose care is traditionally women's work. Irrationally undervaluing women's potential and women's intellect may involve undervaluing children's intellect. As an example of this, consider the following passage from Schopenhauer's 'Essay on Women;' Schopenhauer writes: 'Women are directly fitted for acting as the nurses and teachers of our early childhood by the fact that they are themselves childish, frivolous and short-sighted; in a word, they are big children all their life long – a kind of intermediate stage between the child and the full-grown man, who is man in the strict sense of the word. See how a girl will fondle a child for days together, dance with it and sing to it; and then think what a man, with the best will in the world, could do if he were put in her place' (the implication being that he would not do as well as she; Schopenhauer 1913, 106). Archaic and offensive as this passage may appear today, the undervaluing of children's intellect has remained a contested issue in the recent history of education. Influenced by Piaget, child psychologists in the mid twentieth century and beyond inaccurately taught future teachers that children are incapable of reasoning, incapable of understanding cause-effect connections, and incapable of understanding when another person's ideas and preferences differ from their own. More recent child psychology studies by Alison Gopnik and others have disproved all of these ideas, as was noted earlier; children actually can reason and understand cause-effect connections, we now realize. Thus, Dinnerstein's philosophy has a lot to offer us not only for conceptualizing anti-sexism, but also for opposing lingering, false notions of children's irrationality that are very destructive and inhumane, educationally.

Conclusion: persuasion against gender bias advanced on a basis of self-interest

A particularly challenging question arising from the above may be the question of how to persuade people to be anti-sexist, especially considering that sexism and gender bias may be deeply interwoven in much of what people most value, and into what gives meaning to their lives. Men, and women who want to gain more power in traditionally 'masculine' areas of endeavor, frequently may find that engaging in hyper-competitive activity is desirable for them, and gives meaning and purpose to their lives; therefore, it may be difficult to convince them that hyper-competitive interaction may be a pathological and immature outcome of people being mis- educated to a conventional form of 'masculinity,' and that such consuming competitiveness may be ultimately destructive and contribute to rapacity towards other people and towards the natural environment. And on the 'feminine' side, it may be difficult to convince some women that men can care and nurture just as well as women, and that it would actually be beneficial for women to encourage and support men who are interested in assuming caring and nurturing roles. Women may like feeling that they have a special kind of 'feminine' power to care, to nurture, and to provide sensual pleasure that makes them superior in their own, gendered way; and women may resent being told that they should give up this supposed 'superiority' in order to advance gender equality, abolish traditional 'feminine' and 'masculine' social roles, and help create a better society where rapacity and exploitative behaviors are no longer fueled by traditional, gendered dynamics.

In this situation, what are the best options for persuading people of Dinnerstein's conviction that we must cultivate psychologically deep-rooted gender equality for the good of humanity and of the natural environment?

Perhaps it would be best to argue on a basis of enlightened self-interest. We have arrived at a moment in history where there is more abundant evidence than ever suggesting that one's gender need not constrain one's aspirations intellectually, politically, emotionally, or relationally. Therefore, it may be increasingly likely for people to chafe upon encountering gendered constraints on human aspiration and on human activity. People today may frequently be more eager than in the past to seek solutions to the persistent puzzle of why, although sexism and gender bias are irrational, they still often feel 'right' to people (a puzzle Dinnerstein can help with). This may be particularly true, for instance, in situations where people are wondering why a well-qualified, female candidate for political office was defeated by an obviously less well-qualified, male candidate; or in situations where it is evident that a young man who had interest and potential in the field of early childhood education was dissuaded from actualizing his potential because he received the message that childhood education is not a 'manly' occupation, or that his interest in educating children indicates an interest in sexually assaulting children (a particularly cruel, inhumane, factually unjustified and yet effective deterrent to young men with talents and interests in the area of childcare).

Another, related, persuasive approach is to invite people to consider what an improvement it would be if, with regard to ourselves and with regard to other people with whom we interact, we could have a decreased level of anxiety about the possibility that, due to gender bias, individuals have had experiences of being under-rewarded, under-rated, over-rewarded, and over-rated. Both experiences, of course, are undesirable from purely selfish perspectives. If a male person knows that they have been over-rated and over-rewarded due to sexism (as men and boys often are), then that person naturally comes to feel uncomfortable with their position and distracted from understanding and acting upon the question of what would be the best way for them to live a good life. They may pursue goals that are grandiose instead of worthwhile, they may feel irrationally resentful towards women over whom they have unfairly triumphed, and they may retaliate against women in order to relieve their feelings. Or they may enter into moods of self-disparagement and depression due to a sense of never really being able to know their own value and abilities objectively. And conversely, women, knowing that they are subject to sexist discrimination, may understandably feel resentful and generally hostile towards men and boys.[5]

This situation, which has persisted for so long, surely is one that we all have a rational self-interest interest in ending by destroying gender bias from its deepest roots upwards, despite the fact that those roots may lie in our earliest, childhood experience, and despite the fact that our gendered biases may be involved in some of our most traditionally cherished ways of valuing ourselves and of aspiring to achieve meaningful successes in our lives.

Notes

1. Interestingly, there is extensive historical evidence suggesting that, even while overwhelmingly preferring male power and repeatedly organizing their societies to support excessive enactments of male power, people have simultaneously, to some extent, realized that it is dangerous to do so. In the histories and in the mythological origin stories that various societies have used to explain themselves, there have frequently figured bad, tyrannical kings whose reigns are supplanted by better political leadership. One could point to the ancient Roman tale of tyrannical kings being replaced by that superior governing body, the Roman Republican government; to the traditional Chinese tale of the tyrannical and destructive, Qin dynasty 'First Emperor,' Qin Shi Huang, who was later replaced by better rulers; to the ancient Athenian tale, famously rendered by the playwright Aeschylus in his *Oresteia* trilogy, of the ancient king Agamemnon, the like of whose reign receded and was replaced by the superior governance provided by Athenian democracy; or even to the founding story of the United States as expressed in the U.S. Declaration of Independence, which principally blames the woes of the colonies that became the U.S. not on England's wealthy capitalists and on the Parliament that sought to serve their economic interests through its domineering policies, but rather on the tyrannical King George III.
2. For a discussion of this problem as it relates to misogyny, see Manne (2018).
3. Some recent research suggests that different kinds of empathy are differently related to anti-social, aggressive behavior. A more emotionally mature kind of empathy that is linked to a decrease of anti-social, aggressive behavior may be 'perspective-taking,' which has an intellectual component and is not only emotional. In contrast, the

possibly more immature empathetic responses of distress and fantasy-related empathy (empathy with fictional characters) may actually be linked to increased aggressive, anti-social behaviors. These developments lend credence to the idea that an approach to developing empathy that integrates the heart (stereotypically considered 'feminine') *and* the head (stereotypically considered 'masculine') together may be best, instead of an approach that focuses on educating people to display predominantly 'feminine' qualities or predominantly 'masculine' qualities in response to human needs and human sufferings. For one example of such an approach, see Rosenberg (2003). Rosenberg encourages communication that not only involves people naming their feelings but also considering what deeper needs may be behind the feelings, and what rationally well-considered and respectful action could be taken to address those needs constructively.

4. Further, as an antidote to the way of thinking that unhelpfully and sentimentally over-exalts femininity and women, one can draw upon bell hooks's writing on mothers' abusiveness and their perpetuation of sexism in their education of their own children (Hooks 2000, 73–75). Another useful resource is William Blake's poem "The Mental Traveller", which depicts the cycle across generations of men's violent actions and women's violent actions interacting over time to produce ongoing rapacity and rape (Blake 1988, 483–486). It is necessary not to lose sight of the fact that rapacity and violence are problems arising from 'feminine' behavior as well as from 'masculine' behavior. Also relevant in this connection is experimental evidence of mothers discouraging their young sons from communicating about their experiences of emotion and vulnerability, and evidence of mothers overestimating their crawling baby sons' potential to rise to physical challenges while underestimating their crawling baby daughters' potential to do so (Rivers and Barnett 2011; Mondschein, Adolph, and Tamis-lamonda 2000). Historical and psychological studies examining women's sexist conduct towards other women may also be useful; for a discussion of this see Bynum (2014).
5. Of course, this set of problems comes with a complex subset of related problems including the problem of men being under-rated in their capacity to perform traditionally feminine tasks such as caring and nurturing (tasks that, though crucially socially important, are themselves assigned too little importance in the context of patriarchal society); and the problem of people who have great potential for fine achievement but are not competitively oriented finding that their abilities are under-rated because they are not displaying 'masculine' prowess by defeating other people.

Disclosure statement

No potential conflict of interest was reported by the author.

References

Barton, A. 1974. "Untitled Introductory Essay for *the Taming of the Shrew*." In *The Riverside Shakespeare*, edited by G. B. Evans, 106–109. Boston: Houghton Mifflin.

Blake, W. 1988. *The Complete Poetry & Prose of William Blake*, edited by D. Erdman. New York: Doubleday.

Bynum, G. 2014. "Gender Bias in Women." *Forum on Public Policy Online* 2014: 2. ERIC Number: EJ1059409. -1556-763X.

Dewey, J. 1997. *Democracy and Education: An Introduction to the Philosophy of Education*. New York: Free Press.

Dinnerstein, D. 1999. *The Mermaid and the Minotaur: Sexual Arrangements and Human Malaise*. New York: Other Press.

Fine, C. 2010. *Delusions of Gender: How Our Minds, Society, and Neurosexism Create Difference*. New York: Norton.

Gopnik, A. 2009. *The Philosophical Baby: What Children's Minds Tell Us about Truth, Love, and the Meaning of Life*. New York: Farrar, Straus, and Giroux.

Hickling, A. K., and H. M. Wellman. 2001. "The Emergence of Children's Causal Explanations and Theories." *Developmental Psychology* 37 (5): 668–683. doi:10.1037/0012-1649.37.5.668.

Hodgkinson, T. 2010. *The Idle Parent: Why Laid-back Parents Raise Happier and Healthier Kids*. New York: Penguin.

hooks, b. 2000. *Feminism Is for Everybody: Passionate Politics*. Cambridge, MA: South End Press.

Kushnir, T., and A. Gopnik. 2005. "Young Children Infer Causal Strength from Probabilities and Interventions." *Psychological Science* 16 (9): 678–683. doi:10.1111/j.1467-9280.2005.01595.x.

Kwak, Y., J. W. Payne, A. L. Cohen, and S. A. Huettel. 2015. "The Rational Adolescent: Strategic Information Processing during Decision Making Revealed by Eye Tracking." *Cognitive Development* 36: 20. doi:10.1016/j.cogdev.2015.08.001.

Lucas, C. G., S. Bridgers, T. L. Griffiths, and A. Gopnik. 2014. "When Children are Better (Or at Least More Open-Minded) Learners than Adults: Developmental Differences in Learning the Forms of Causal Relationships." *Cognition* 131 (2): 284–299. doi:10.1016/j.cognition.2013.12.010.

Magary, D. 2013. *Someone Could Get Hurt: A Memoir of Twenty-First Century Parenthood*. New York: Penguin.

Manne, K. 2018. *Down Girl: The Logic of Misogyny*. New York: Oxford University Press.

Mondschein, E. R., K. E. Adolph, and C. S. Tamis-lamonda. 2000. "GenderBias in Mothers' Expectations about Infant Crawling." *Journal of Experimental Child Psychology* 77 (4): 304–316. doi:10.1006/jecp.2000.2597.

Rivers, C., and R. C. Barnett. 2011. *The Truth about Girls and Boys: Challenging Toxic Stereotypes about Our Children*. New York: Columbia University Press.

Rosenberg, M. B. 2003. *Nonviolent Communication: A Language of Life*. Encinitas, CA: PuddleDancer Press.

Saini, A. 2018. *Inferior: How the New Science Got Women Wrong– And the New Research That's Rewriting the Story. Reprint*. Boston: Beacon.

Schopenhauer, A. 1913. *Studies in Pessimism: A Series of Essays*. edited by. T. B. Saunders. London: George Allen & Company. As reproduced online at https://en.wikisource.org/wiki/Studies_in_Pessimism/On_Women

Educating *in* and for *uncertainty*: Climate science, human evolution and the legacy of Arne Naess as guidance for ecological practice

Margarita García-Notario

ABSTRACT

This paper reflects on how the issue of climate change and the general state of our planet is, among other causes, a main factor in the paralyzing divisions ailing Western societies. This situation, while unsettling to democracies, is promoting a kind of education *in* and *through* fear and I question if education can succeed under these circumstances without becoming indoctrination. This paper does not try to diminish the urgency and the importance of current environmental problems but rather expands today´s perspectives and incorporates research in more constructive ways of thinking and doing. I use scientific contributions in climatology, evolution, environmental conservation, economics, and neuroscience to bring new light to today's investigations about the human and the non-human world. Finally, I propose Deep Ecology's principles of *deep questioning, deep experience* and *deep commitment*, as a guide for new educational and ecological practices.

Introduction. distressing fears

We seldom see anybody who is not uneasy or afraid to live.[1] Ralph Waldo Emerson I have noticed as a long-time teacher and researcher how much more fearful our youth are today about the future. Many colleagues share this impression: students are pessimistic and many (too many) suffer from anxiety and depression. Authors Jonathan Haidt and Greg Lukianoff cite a study published in the Journal of the American Medical Association showing an alarming spike in suicide rates between 2011 and 2015 among teenagers.[2] Haidt and Lukianoff blame principally the pervasive use of social media, but many college and high school teachers also report students' extreme anxiety over the ecological state of our planet.

The young have taken to the streets. They join large and vociferous movements demanding prompt government action to address and ameliorate the effects of climate change. Many are sure that their generation is doomed by today's environmental troubles, even if they lack rigorous scientific knowledge and instead base their conclusions on propagandistic slogans. Greta Thunberg has become a new star in the environmental movement and her message resonates with adults and youth. In a recent speech, Thunberg said, 'Adults keep saying we owe it to the young people, to give them hope. (. . .) But I don't want your hope. I do not want you to be hopeful. I want you to panic. I want you to feel the fear I feel every day. (. . .) I want you to act as you would in a crisis. I want you to act as if the house is on fire, because it is. (. . .) This is the biggest crisis humanity has ever faced.'[3]

Climate change is certainly an important challenge for humans, but is it the *biggest crisis humanity has ever faced*? A detailed scientific answer to that question is beyond the scope of this paper, and I believe that we must confront bad news and not ignore or disguise it just to protect others from negative emotional responses. However, when an issue starts looking more like a political flag than a dispassionate examination of the scientific evidence and its effects are harming our youth, we teachers and philosophers of education must ask (more) questions. It is our stock and trade to search for truth and meaning and better perspectives. A perception of threat keeps us in a physiological defensive *flight or fight* state and blinds us from seeing the value of contrarian or different arguments. When the premises of the issues discussed rest on fears and prejudices, there is little progress or possibility of fruitful dialogue. We have the responsibility to present to students reliable scientific information that shows both the science that may suit our beliefs and preferences but also that which questions them. Like Socrates, we need to act as a gadfly on a horse [4] as we try to awaken students to reflect on positions and biases that could distract them from their true human interests in truth, justice, and wisdom. In the last few years I have followed closely these environmental challenges and proposed solutions. Some of my findings were very inspiring and gave me and my students renewed optimism in relation to our environmental concerns. In the next section, I present some recent and exciting research on climate change that highlights the relationship between climate and human evolution.

Climate change: friend or foe?

Rick Potts is a paleoanthropologist at the Smithsonian Institute. His studies of layers of rocks and sediments in West Africa made him question the traditional idea that it was the savannah, the grassy plain with few trees that had been human evolution's driving force. Instead, he and his colleagues realized that *variability* itself was the driving force of human evolution, and that our ancestors became our ancestors by adapting to change itself. This is a simple but revolutionary idea: *human evolution is nature's experiment with versatility*. We are not adapted to any one particular environment or climate, but to many. According to these studies, humans are creatures of climate change.[4]

Early human fossils found in the African continent go back six million years and reveal ancient creatures walking upright on the ground, although they also spent time in the trees using them to sleep in and to escape predators. These creatures, our human ancestors, were at home in a combination of two worlds: the ape world and the early human world. DNA studies locate between five and seven million years ago the moment when humans and chimpanzees shared a common ancestor: Sahelanthropus Tchadensis. And they place around two million years ago our first human ancestor: the toolmaker or Homo Habilis. Africa was home to many varied human- like species and when scientists study their environment they find that climate instability pressured them all. In a period around 2,000,000 years, 'in the evolutionary blink of an eye, punishing drought periods alternated with storms and monsoons. Rivers and forests sprang up, then turned to dry grassland. (. . .) It was these periods of climate instability that seem to be the main clue to understanding the evolutionary leap from small bipedal apes to the larger-brain toolmaker Homo Habilis.[. . .] Today, climate change seems to threaten our survival, but it may have held the keys to the astonishing story of how we became who we are, because it didn't stop 2,000,000 years ago. These dramatic climatic upheavals would continue for another million and a half years, propelling our ancestors down a road leading, ultimately, to the smartest creature the world has ever known.'[5]

Theory and the fossil record also demonstrate in the science of conservation biology that evolutionary creativity is essential for long-term survival in a changing physical context. 'Ecotypes or ecological races arise in response to novel challenges, both biotic and abiotic. The future of endangered species is likely to depend on such processes. Failure to appreciate this is the single biggest flaw in the *climatic envelope* or *niche modeling* approach to conservation biology.'[6]

These studies are mindboggling and truly inspiring. Amid logical fears of what climate change may wrought, I welcome these revelations of how we became who we are, as it empowers me to recognize that since we have survived such immense challenges we will likely survive those to come.

Rambunctious gardens and skeptical environmentalists

From the start of the modern environmental movement, arguments over the balance between humans and nature have polarized: anthropo-centrists versus eco-centrists. The effects of climate change have intensified the arguments. Many conservationists would like to keep humans at a distance from natural places, claiming that nature is only in balance without man. But is that accurate? Those appeals for *balance* conflict with one of humans' earliest observations, articulated by Heraclitus: *The only constant in nature is change itself.*

Writer Emma Marris claims that if we define *nature* as that which is untouched by humans, then we won't have any left. She proposes a new definition of nature -one that includes not only pristine wilderness but also the untended patches of plants growing in urban spaces- and encourages us to bring our children out to touch and tinker with it, so that one day they might love and protect it.[7] In her book Rambunctious Garden: Saving Nature in a Post-Wild World[8],

Marris questions the efforts of environmental conservation approaches to safeguard wild areas, since this prevents a fuller relationship with nature. Humans have changed the landscapes they inhabit since prehistory, and climate change means even the remotest places now bear the fingerprints of humanity. She and others claim that we need different approaches such as rewilding, assisted migration, and the embrace of so-called novel ecosystems.[9]

Some have put Emma Marris in a group called Modernist Greens. This group's soldiers don't dispute the ecological tumult associated with the Anthropocene[10]. But they look at the world with today's eyes and ask that we 'reconcile the needs of people with the needs of nature [... and] craft a new vision of a planet in which nature – forests, wetlands, diverse species, and other ancient ecosystems – exist amid a wide variety of modern, human landscapes.'[11] This approach helps parents of young children to catch their breath. Expanding the definition of nature gives us permission to go outside and offer children a friendlier way to exist in the world, indeed, *with* the world.

Bjorn Lomborg is a Danish economist who acknowledges the gravity of global warming,[12] but presents scientific and economic evidence to show that today's alarmism is misguided and keeps us from thinking intelligently about solutions. Lomborg and the Environmental Assessment Institute founded the Copenhagen Consensus in 2002, which seeks to establish priorities for advancing global welfare using methodologies based on the theory of welfare economics. Lomborg is concerned about the enormous cost attached to many of the proposed solutions to stop global warming and claims that many of these are often based on emotions rather than rigorous science. He warns that most proposed solutions would have little impact on the world's temperature for hundreds of years. Rather than starting with the most radical procedures, Lomborg argues that 'we should first focus our resources on more immediate concerns, such as fighting malaria and HIV/AIDS and assuring and maintaining a safe, fresh water supply – which can be addressed at a fraction of the cost and save millions of lives within our lifetime.'[13] Lomborg's work opens a second level of thinking on global warming, one that argues panic is neither warranted nor a constructive place from which to deal with any of humanity's problems, let alone global warming. He warns, for instance, 'abandoning fossil fuels as quickly as possible, as many environmental activists demand, would slow the growth that has lifted billions of people out of poverty.'14

In spite of the general impression that things are terrible, Lomborg's research (together with others like Stephen Pinker, etc.[15]), shows that our world is mostly getting better for humans: global life expectancy has more than doubled since 1900[16] and is now above 70 years; health inequality has declined massively. The world is more literate;[17] child labor is decreasing;[18] violence is down; people are better off economically and poverty has dropped; more people have access to improved water sources; and the risk of death by air pollution has declined substantially ... [19] There is much good news for both humans and the planet and both young and old people need to realize this. We should not drown in an ocean of threats so large that progress seems lost.

Whether Lomborg's claims are a better way for environmental solutions or not, I believe that the main challenge for educators (as opposed to indoctrinators) is to review *all* contributions and evidence with the intention to learn and improve, not to serve a more or less political status quo. It is quite disheartening to acknowledge that it has become fashionable to label an opponent as *anti-science* or as *science denier*, particularly in debates about the climate and agricultural biotechnology.[20] It would enrich all of us in academia and enliven our debates if we took to heart Alice Dreger's advice at the end of her book Galileo's Middle Finger: 'If you must criticize scholars whose work challenges yours, do so on the evidence, not by poisoning the land on which we all live.'[21]

The human need for safety. the polyvagal theory[22]

Feeling safe is a very strong manifestation of our survival instincts and a *sine qua non* for living beings to progress and thrive. Dr. Stephen Porges' neurobiological discoveries through his Polyvagal theory provide new insights to the human necessity for safety. They give educators a refreshing perspective that enables us to understand the why and how of human need for contact and connection with each other. Moreover, it shows that, when we do not connect with each other, our physical and mental health suffer.

According to Porges many think that being safe means the absence of threats. But the mere removal of threats will not make humans safe. Feeling safe comes from having good social interactions, which comes from living in a trustworthy social system.

Evolution shows that we have a biological imperative to survive and in mammals and humans this imperative is realized by connecting with others. From this perspective, social polarization threatens the biological imperative, and according to Porges, today's situation (meaning the strong social and political divisions in a lot of the Western world) is evolutionarily unsustainable. When we detect danger or feel under pressure, we shift biological states and, as a consequence, we shift our bias: the way we see the world and others changes, and we get into a defensive state.[23] Polarization overrides dialogue and reciprocity; both are only possible when there are enough cues for safety.

Understanding our neurobiology can help us realize that not all of our behaviors are intentional and that we co-regulate our health and behaviors through social interactions. Let me explain in more detail with the Polyvagal theory.

The so-called polyvagal system gets its name from the *vagus* nerve and has to do with the circuits regulated at the brainstem that send information through our nerves in a bidirectional way: up to the brain and down to the body. It regulates our human experiences of trust and safety. Humans need to feel safe to develop their curiosity and to be open to new information and experiences. When we feel unsafe, we start to see everything unfamiliar as a threat and this makes us unable to take in new information.

By bringing together psychology, neuroscience and evolutionary biology, the polyvagal theory gives a new understanding of the Autonomic Nervous System (ANS), by showing that it consists of 'three neural circuits that form a phylogenetically ordered response hierarchy that regulates behavioral and physiological adaptations to safe, dangerous and life-threatening environments.'[24] David Fuller[25] summarizes it as: 'a theory (...) that explains how our body and mind are intimately connected to our facial expressions, voice and social engagement system and therefore, to others.'[26]

The Polyvagal theory demonstrates that the regulation of our bodily functions follows an evolutionary trajectory and we find that the *vagus* system contains the different stages in the origins of life. Our most primitive autonomic system is the one we share with the oldest planetary living beings. It is the *older dorsal system* that connects directly to the gut, and its purpose is to maintain metabolic resources: oxygen and food regulation; if there is neither enough oxygen nor food, the system shuts down to conserve energy. This part of the *vagus* appeared first in evolution and it is non-myelinated. Physiologically, this part of the vagus, also known as parasympathetic system, is located in the brainstem and it appears simultaneously with the first biological instinct: the sexual instinct, which caused an explosion in the variety and in the possibilities of novelty in our planet. 'This biological innovation (sex) vastly increased the possibilities for generating the variability, diversity and complexity of species that has produced the richness that is the wonder of life in this planet'.[27]

The next *vagus* area is the sympathetic system, which provides mobilization and is controlled by the limbic part of the brain that developed with the arrival of the reptiles. It is commonly known as the *flight or fight* response. It is at this moment when the next instinct, the 'parental instinct' appears. Through it 'organisms are instinctively impelled to protect and nurture their offspring.'[28]

The newer *vagus* area is the one that develops with the arrival of mammals and brings the third instinct: the social instinct. This originates physiologically in the newest and youngest limbic area exclusive to mammals. Around 216 million years ago, the arrival of mammals brought an unusual characteristic: sociability. The new limbic area adds three neurological abilities: 1- an intimate and prolonged relationship between the newborn mammal and the mother (and sometimes the father). 2- The ability to maintain audio-vocal contact with the mother/father/group, and origin of the *separation cry*: a new behavior that attracts the attention and thus protection from the mother/father/group. 3- The ability to get pleasure through play.

This part of the *vagus* is different because it is covered with a fatty substance (myelin) which makes the communication with the brain much faster. The mission of this part of the *vagus* is to calm down and organize the older *vagus*: it connects with a different part of the brain and it does not go to the gut, but to the heart and to the facial and ear muscles: it brings our feelings to our face. When we feel safe, our heart makes our facial muscles relax, our eyes get crinkles and the orbital muscle connects with the middle ear muscle that gets tense and is then able to hear the higher pitch of the human voice. We can then quiet the lower predatory tones, causing us to relax, feel safe, and engage with others. When we feel in danger our brain has difficulty finding safety and we are in fight-flight mode, a different physiological state called mobilization; if we can fight or flee, we have taken action to deal with the fear. In mobilization, we are more likely to misread other people's cues. The social interactional behavior is a neural exercise which uses newer mammalian structures aimed at inhibiting more primitive systems.

Through gentle and appropriate touch, tone of voice, and smiles, people calm one another and making them feel safe.

The Polyvagal theory provides many insights with respect to our interactions with others and helps us see that many of the evaluations and actions that the nervous system makes are unconscious. Porges calls that evaluation *neuroception*: a knowledge that our body gains by picking up the external cues of safety. The moment that we perceive threats through neuroception, we lose our social engagement. Mammals and humans, with their more developed neurological system use their newest system (social engagement) first to co-regulate. However, when the newer (social) system fails, either by lack of appropriate social engagement or by neurological injury, the older systems of defense (fight or flight) are disinhibited. (Antonio Damasio explained this in his book Descartes' Error, by showing how the Jacksonian principle of dissolution worked in patients with neurological injuries). Newer systems override older systems, but if the perception of threat persists, the older system takes over.[29]

This new science has been revolutionary in therapeutic treatments because the therapists have gotten a much broader and deeper knowledge about how to read their patients' facial and body signs of anxiety, trauma, etc. Multiple techniques have been developed to help patients heal by restoring biological and psychological safety. This physiology lesson teaches us that the behaviors promoted by the newer vagus such as making eye contact, vocalizing with inflection and rhythm, observing our interlocutor's facial expressions and listening better, become great tools to increase learning and to combat social polarization through the promotion of mutual understanding and cooperation.

In 2015, in a presentation at the ECER Conference in Budapest[30], Sharon Todd articulated brilliantly the problems and errors of the *Euro Vision*: the document that drafted the educational policy guidelines for the European Union. She claimed that it is a mistake to make education the solution for social uncertainty and expect that it can assuage the problems of the economy. She quoted Hannah Arendt to warn that this view of education 'puts the responsibility of solving and fixing the worlds' problems on the youth who have had no hand in causing these problems and suggests an inability to tolerate uncertainty itself.'[31] 'Humans may crave absolute certainty -says Carl Sagan-; they may aspire to it; they may pretend ... to have attained it. But the history of science – by far the most successful claim to knowledge accessible to humans – teaches that the most we can hope for is successive improvement in our understanding, (...) but with the proviso that absolute certainty will always elude us.'[32]

Uncertainty is the most *certain* quality of human life, but the level of uncertainty that allows human life to thrive has to be moderate and temporary, or we humans would spend our lives in defensive postures, which harms our bodies and keeps us from productive and insightful thinking and adaptation. It is necessary for the adults in the room to recognize when the input that the youth receive is too much and shuts them down. Adults have to act with the *old* wisdom characterized by truth and prudence or moderation.

Towards a new ecological practice. some inspiration from Arne Naess

Is there a different way through which the new responsibility that our generation feels towards the Earth could unite us in finding ways to act that are better for both the environment and humans?

In this final section, I would like to talk about the Norwegian philosopher, Arne Naess, founder of Deep Ecology, one of the earliest eco-philosophies that developed after scientific warnings about human culpability in ecological crises. He coined Deep Ecology's name at a conference in Bucharest in 1972, as an alternative or different orientation inside environmentalism to **superficial ecology**, which was a movement focused on the fight against pollution and the destruction of natural resources and whose main goal was the health and the wealth of people in developed countries. Without diminishing the importance of those concerns, **Deep Ecology** sought to be a movement that transcended the *superficial* vision, by raising important questions about our understanding of nature, of humans, and of the relationship between the two. Although Deep Ecology has been defined in various ways, I believe that Stephan Harding summarized its essence best with the slogan 'Deep Questioning, Deep Experience, Deep Commitment.' I find this slogan most appropriate to educational approaches that acknowledge the role of uncertainty in human life, and that use human tools to help us live better in the world that hosts us and whose destruction we would share.

Deep questioning

Naess claimed that humans act from a set of fundamental premises or beliefs, independently of their ability to verbalize them. Those premises give meaning to their lives. Upon recognizing the ecological crisis, Naess said that something *deeply ours* is damaged as well. The deep questioning that he encourages is active, encompasses the whole person, and is positive; it is not guided by a desire to undo what others have done. It is a kind of comparative and embracing questioning. The belief that *everything hangs together*, that all life is related, is the main propeller for new questions: what else is there that I don't see? What can the perceptions and knowledge of others teach me? This questioning process should always be open, and the conclusions should remain open to future corrections or modifications.[33]

A most fecund educational approach results from a way of living that is guided by an attitude of deep questioning and an ecological context that enriches it, because it forces us to recognize our relationship with the other (non-human) beings and it invites us to have an openness towards change, correction, and life, which is always expanding. This deep questioning also requires for Naess that reason and feelings do not act separately. We need to acknowledge our feelings and to investigate them; we must learn their lessons and use them to better ourselves and others (including the non-human world). Our feelings will help us engage in deeper experiences and, through them, invite deeper commitments.

Deep experience

The key role that feelings and emotions played in Naess's thought is the content of his last book, Life's Philosophy. Baruch Spinoza is a key philosophical influence on Arne Naess, who refers to Spinoza as (maybe) the least original among the modern philosophers, but the one who has taken emotions the most seriously and the one who has reflected on them with the most depth. For Spinoza, 'the path to freedom is paved by emotions'; and humans can only progress and thrive through them and apathy (the absence of pathos: feeling, suffering) equals spiritual death.

Spinoza distinguishes between active and passive emotions in the same way that he distinguishes between the terms *action* (from the Latin: *actio-actionis*) and *activity* (from the Latin: *activitas-activitatis*). The true human doing (acting) is represented by the term *action* because it implies (requires) improvement. It is the kind of doing that affects us and transforms us as human beings. (In my opinion, we could call it *virtue*, as the good habit that guarantees our good behavior: it improves us). By contrast, *activity* is an external doing of our actions: it ends outside ourselves. Humans, according to Spinoza, need to be sure that they invest in the actions that transform them, that make them better human beings. It is a risk and a temptation to invest and engage ourselves in multiple activities (busyness) that have no good or permanent effects on us.

When Spinoza separates active from passive emotions, he gives little value to the latter, while Naess disagrees about the secondary role given to passive emotions. Naess claims that these can and should be used in a constructive way and that they have a very complex and indispensable role in human life. It is through negative emotions that we change course, for example. Emotions help us to act for passion and passionately. In today's context of the ecological crisis, emotions such as ire, fear, and deception can be great assets to force ourselves to work towards change. For both Spinoza and Naess, feelings are the main source of motivation and the most important tool to change what we are doing when we want to feel different.[34] We only progress in the things that are essential to humanity when we act through passion.[35] For Naess, much of an individual's emotional maturation happens by transforming negative feelings into positive ones. What insightful guidance this is for today's educators and what a productive way to transform our despondence towards environmental damage into an effort to restore ourselves and nature.

Deep commitment

Theodore Roszak warned in an interview in 1998[36] that the proliferation of bad news about the environment can paralyze, which in many ways inhibits creative change. The enumeration of one disaster after another sends us in search of the guilty ones but in a counterproductive way. We may be overdoing accusations and threats and creating wide-spread panic. When this happened in the early discussions about the environmental crisis and people couldn't take it anymore, they responded by making fun of the warnings and denying and evading the problem. Today we may be at the opposite side, making this crisis a black hole that eats up the good and the bad.

We need to aim towards compromise, towards working together even if in imperfect ways. 'Deep commitment is the result of combining deep experience with deep questioning, – writes Stephan Harding. – When an ecological worldview is well developed, people act from their whole personality, giving rise to tremendous energy and commitment. Such actions are peaceful and democratic and will lead towards ecological sustainability. This leads to 'extending care to humans and deepening care for non-humans'.[37] I am persuaded that this reflection is true and achievable. And I believe that we can obtain measurable and lasting progress, both in our environmental situation and in our democracies so embattled today by persistent ideological divisions.

Final thoughts

This paper is an effort to articulate my concerns about today's educational messages and the way that they instill dread and doom in our youth; I wonder if it may be losing a sense of proportion and defeating its own purposes. I have felt that many of my students have overcome their fears by learning about our evolutionary history and the factors that have contributed to make us who we are today. I have confirmed that bringing controversial voices to the class, has enriched the students' discussions greatly and has encouraged them to review their previously unquestioned theses.

We humans are messy, and we struggle to balance our minds with our hearts.

Nevertheless, it is good to know about the physiological mechanisms that support our behavior and to learn strategies to prevent reverting to actions guided by fear instead of sound emotions reason. In our uncertain future, it is important to be more mindful in the present. It is my hope that the examination of the biological imperative that we connect with others will encourage all of us to promote social engagement behaviors to improve learning and work environments, to think with serenity, to use our creativity in productive ways, and to stop avoiding the ideas that challenge our cozy biases. We must give them a chance to either correct us or confirm that we are on the right path.

[1] Ralph, Emerson. 'Prose Works. v.2. – Full View | Hathi Trust Digital Library *Hathi Trust*.

[2] Lukianoff, Greg, and Jonathan Haidt. *The Coddling of the American Mind.* (NY, Penguin, 2018).

[3] Greta Thunberg. Speech at the World Economic Forum Annual Meeting (2019).

[4] Potts, Richard, and Christopher Sloan. *What Does It Mean to Be Human?* (Naess and Haukeland 2020).

[5] 'Becoming Human Part 1 | NOVA | PBS.'

[6] Arthur, Shapiro. "Professor Arthur Shapiro's Review of Emma Marris' Rambunctious Garden

– Conservation Sense and Nonsense."

[7] Marris, Emma. 'Emma Marris: Nature Is Everywhere – We Just Need to Learn to See It | TED Talk.'

[8] Marris, Emma. *Rambunctious Garden*. (Bloomsbury Publishing USA, 2013).

[9] Ibidem.

[10] The Anthropocene is a growing scientific consensus that the contemporary human footprint – our cities, suburban sprawl, dams, agriculture, greenhouse gases, etc. – has so massively transformed the planet as to usher in a new geological epoch.

[11] Kloor, Keith. (12 April 2012) *The Great Schism in the Environmental Movement. Can modern greens loosen nature's grip on environmentalism?*

[12] See Lomborg's referenced books in the bibliography.

[13] Udgave, Dansk. 'Bjørn Lomborg: Cool It.'

[14] Lomborg, Bjørn. 'The Danger of Climate Doomsayers by Bjørn Lomborg – Project Syndicate.'

[15] Pinker, Steven. *Enlightenment Now*. (Penguin Books, 2019).

[16] Roser, Max, et al. 'Life Expectancy – Our World in Data.'

[17] Roser, Max, and Esteban Ortiz-Espina. 'Literacy – Our World in Data.'

[18] Ortiz-Espina, Esteban, and Max Roser. 'Child Labor – Our World in Data.'

[19] Lomborg, Bjorn. See note 14.

[20] Jr, Roger Pielke. 'Why Discrediting Controversial Academics Such as Bjørn Lomborg Damages Science | Environment | The Guardian.'

[21] Dreger, Alice. *Galileo's Middle Finger*. (Dreger 2016).

[22] Porges, Stephen. 'The Polyvagal Theory: New Insights into Adaptive Reactions of the Autonomic Nervous System.'

See also: Chitty, John. 'The Triune Autonomic Nervous System.'

[23] Ibidem.

[24] Ibidem.

[25] Founder of the British Podcast *Rebel Wisdom*.

[26] Wisdom, Rebel. *Dr Stephen Porges: The Neuroscience of Polarisation (Pt 2 of 4)*. YouTube, 15 August 2019, https://www.youtube.com/watch?v = vmkG5I7CaGw.

[27] Margulis, Lynn. *Symbiosis in Cell Evolution*. W H Freeman & Company, 1981. Quoted in Loye, David. *Darwin's Lost Theory*, p. 78.

[28] Loye, D. Op. Cit. P. 81.

[29] Binz, Kevin. 'Jacksonian Dissolution | Fewer Lacunae.'

[30] Sharon, Todd. 'Sharon Todd | EERA.'

[31] Ibidem.

[32] Sagan, Carl. *The Demon-Haunted World*. 1st ed., (Sagan 2011), 28.

[33] This attitude was frustrating to some of Naess' followers. Witoszek, Nina, and Andrew Brennan. *Philosophical Dialogues*. (Witoszek and Brennan 1999), 35.

[34] Naess, Arne, and Per Ingvar Haukeland. *Life's Philosophy*. (Naess and Haukeland 2020), 78.

[35] Baruch Spinoza, quoted in Naess, *Life's Philosophy*. p. 9.

[36] Toms, Michaels. 'Deep Ecology for the 21st Century. .' *New Dimensions Radio*, New Dimensions Radio, 1998, http://www.newdimensions.org.

[37] Harding, Stephan. 'What Is Deep Ecology? | Schumacher College.' *Schumacher College*, 28 September 2007, https://www.schumachercollege.org.uk/learning-resources/what-is-deep-ecology.

Disclosure statement

No potential conflict of interest was reported by the author.

References

2019, http://hdl.handle.net/2027/njp.32101013187024

"Becoming Human Part 1 | NOVA | PBS." *PBS: Public Broadcasting Service*, NOVA, http://www.pbs.org/wgbh/nova/video/becoming-human-part-1 Accessed 29 October 2020.

Binz, K. 2019. "Jacksonian Dissolution | Fewer Lacunae." *Fewer Lacunae* 22 (Apr). http://kevinbinz.com/tag/jacksonian-dissolution/

Chitty, J. "The Triune Autonomic Nervous System." Colorado School of Energy Studies, Accessed 4 February 2019, https://www.energyschool.com/.https://static1.squarespace.com/static/5be365255cfd79dc41e5dc90/t/5c05f753032be40de96a2ce9/1543894895695/triunewww14.pdf

Dreger, Alice Domurat. Galileo's Middle Finger: Heretics, Activists, and the Search for Justice in Science. New York: Penguin Press, 2015.

Emerson, R. W. "Prose Works. V.2. - Full View | HathiTrust Digital Library | HathiTrust Digital Library." *HathiTrust*, Boston, J. R. Osgood and Co.,1872., Accessed December 28.

Fuller, D. "Dr Stephen Porges: The Neuroscience of Polarisation (Pt 2 of 4)." *Rebel Wisdom*, YouTube, 15 August 2019, https://www.youtube.com/watch?v=vmkG5l7CaGw

Harding, S. 2007. "What Is Deep Ecology?" *Schumacher College* 28. September:https://www.schumachercollege.org.uk/learning-resources/what-is-deep-ecology

https://slate.com/technology/2012/12/modern-green-movement-eco-pragmatists-are-challenging-traditional-environmentalists-over-scope-of-nature.html

Jr, Roger., P. "Why Discrediting Controversial Academics Such as Bjørn Lomborg Damages Science | Environment | The Guardian." *The Guardian*, The Guardian, 23 April 2015, http://www.theguardian.com/science/political-science/2015/apr/23/playing-the-ball-not-the-man

Kloor, K. "The Great Schism in the Environmental Movement. Can Modern Greens LoosenQ11 Nature's Grip on Environmentalism?". *Slate.Com* , https://slate.com/technology/2012/12/modern-green-movement-eco-pragmatists-are-challenging-traditional-environmentalists-over-scope-of-nature.html, Accessed 10 August 2019

Lomborg, B. "The Danger of Climate Doomsayers by Bjørn Lomborg – Project Syndicate." *Project Syndicate*, https://www.facebook.com/projectsyndicate, Accessed 19 August 2019, http://www.project-syndicate.org/commentary/climate-change-fear-wrong-policies-by-bjorn-lomborg-2019-08

Lomborg, B. 2020. *False Alarm*. Basic Books

Loye, D. *Darwin's Lost Theory*. David Loye, 2007.

Lukianoff, G., and J. Haidt. 2019. *The Coddling of the American Mind: How Good Intentions and Bad Ideas are Setting up a Generation for Failure*. New York: Penguin Press.

Margulis, L. 1981. *Symbiosis in Cell Evolution: Life and Its Environment on the Early Earth*. San Francisco: W.H. Freeman.

Marris, E. "Emma Marris: Nature Is Everywhere – We Just Need to Learn to See It | TED Talk." *TED: Ideas Worth Spreading*, Accessed 10 November 2019a, http://www.ted.com/talks/emma_marris_nature_is_everywhere_we_just_need_to_learn_to_see_it?language=jv

Marris, E. "Rambunctious Garden: Saving Nature in a Post-Wild World: Emma Marris: Bloomsbury USA." *Bloomsbury Publishing*, 7 Accessed April 2019b, https://www.bloomsbury.com/us/rambunctious-garden-9781608194544/

Naess, A., and P. I. Haukeland. *Life's Philosophy*. University of Georgia Press, 2010. Ortiz-Espina, Esteban, and Max Roser. "Child Labor." *Our World in Data*, http://ourworldindata.org/child-laborAccessed 29 October 2020.

Pinker, S. 2019. *Enlightenment Now*. Penguin Books.

Porges, S. 2018. "The Polyvagal Theory: New Insights into Adaptive Reactions of the Autonomic Nervous System." *PubMed Central (PMC)* 3 (May). https://www.ncbi.nlm.nih.gov/pmc/articles/PMC3108032/

Potts, R., and C. Sloan. 2010. *What Does It Mean to Be Human?* National Geographic Books. Washington, D.C.

Roser, M., et al. 2019. "Life Expectancy." *Our World in Data* 10 (Sept). http://ourworldindata.org/life-expectancy

Roser, M., and E. Ortiz-Espina. "Literacy." *Our World in Data*, http://ourworldindata.org/literacy Accessed 29 October 2020.

Sagan, C. 2011. *The Demon-Haunted World*. 1st ed. ed. Ballantine Books. New York City.

Shapiro, A. "Professor Arthur Shapiro's Review of Emma Marris' Rambunctious Garden – Conservation Sense and Nonsense." *Conservation Sense and Nonsense*, https://www.facebook.com/WordPresscom, 2 April 2013, http://milliontrees.me/2013/04/02/professor-arthur-shapiros-review-of-emma-marris-rambunctious-garden/

Sharon, T. "Sharon Todd | EERA." *EERA*, Accessed 11 november 2019, http://eera-ecer.de/previous-ecers/ecer-2015-budapest/programme-central-events/keynote-speakers/sharon-todd/

Toms, M. "Deep Ecology for the 21st Century". *New Dimensions Radio*, New Dimensions Radio, 1998, http://www.newdimensions.org

Udgave, D. 2019. "Bjørn Lomborg: Cool It." *Klimadebat.Dk - Dansk Forum Om Klima Og Energi* 30. January:http://www.klimadebat.dk/bjoern-lomborg-cool-it-b33.php

Witoszek, N., and A. Brennan. 1999. *Philosophical Dialogues*. Rowman & Littlefield. Washington, DC.

Co-creation in the commonwealth: Understanding right relationship in place

Mark Beatham

ABSTRACT
Could public education as a cultural institution promote the commonwealth? ('Commonwealth,' from The Oxford English Dictionary as, 'public welfare; general good or advantage;' and 'to a body or a number of persons united by some common interest.') This paper argues proper education enfranchises the young through proper relationships to place, past and present, culture and creation, life, and work. Wendell Berry is the principal guide and standard in describing and considering proper relationships in the commonwealth and their consequences. Other major authors include Wes Jackson, Gustavo Esteva, Vine Deloria, Alan Watts, Matthew Crawford, Roger Scruton, Nablan and Trimble, Alison Gopnik. Proper relationships, defined essentially in terms of proper scale, play, and learning, preserve the means of co-creation between human culture and creation. They are measured by health, durability, beauty, harmony, resilience, fecundity, and wellbeing.

We came as a poor people to a *seemingly* empty land that was rich in resources. We built our institutions with that perception of reality. Our political institutions, our educational institutions, our economic institutions – all build on that perception of reality. In our time we have become rich people in an increasingly poor land that is filling up, and the institutions don't hold (Jackson 2011, 116–117).

> How shall our children live in a world from which first the spirit, then history, and finally nature have fled, leaving only the mindless mechanics of process and chance: Will any place exist for a humane art in a society from which the last trace of reverence — any reverence — has been rubbed out (Hayden Carruth, in Berry 2010a, 69)?

Early in the semester, I take my education students outside on a walking tour of the campus and town. Students are bewildered at first, but they quickly gain their feet and senses. Our state college was built in the center of town. From our public building, we cross the street and we are in the 1960s private residences.

By the end, in a mile to downtown and the 1780s, we have zig-zagged between campus and town, looking at what humans have made of where they live and of how what they valued and believed can be found what they have left. We see and consider: how the town grew from a Revolutionary War plot at the mouth of the Saranac River to become a center of commerce for the mining and timber industries; how the oldest buildings, in their Greek and Romans references,[1] still embodied in their designs the promotion of civic virtues; how that civic mindedness led to the first private school, the Plattsburgh Academy, which was quickly shuttered to serve as an armory in the fight the British in the War of 1812; how later the first public high school was built in 1875 on the same site and replaced in 1940 with the WPA project for a public library; how the private built spaces, going west and away from the river, embodied changing concepts of public and private interests; and how the farther you went from the old downtown and in the space of a only a few blocks one could see a literal and figurative move from more civic virtues toward more privatized and exclusive beliefs, enabled by cars and air conditioning. The newer private buildings had no relation to one another or to the public. They were stand-alone tokens of the new privacy. The college campus expressed this especially with its brutalist buildings built in the building boom for Baby Boomers, when energy was cheap (and assumed to be ever so) and the fetish for forms abstract and jutting evincing bland indifference. The college built arrogantly, cheap and ugly, ignoring lessons learned from living in an actual place with actual conditions and traditions.[2] This from a public institution with little sense of the public.

Some built environments show awareness of place – in high pitched roofs and stacked floors and public and private and commercial buildings with 'public faces.' But more show that neighbors present and past have been turning steadily from attention to place and civics by using fossil fuels to dodge acknowledgement of what is essential. Their settlements have not settled properly, depending as they have and do on long supply lines to other depletors and on a drawdown of civic capital. We still have not figured out how to live sustainably and peaceably with local creation. That task falls to us.

The tour is a great way to walk into and through important themes for an education foundations course. The tour and the course consider central questions of a life worth living in particular and of the ways that we as a social species can help one another flourish by preserving the means in the commonwealth with education as one of many critical means. In the following, I discuss the essential qualities of these proper relationships – attending to proper scale, work, play, and learning – to preserve the means of co-creation between human culture and creation. Sustainability requires a proper relationship between culture and nature. Wendell Berry (2014) states this plainly:

> The problem of sustainability is simple enough to state. It requires that the fertility cycle of birth, growth, maturity, death, and decay — what Albert Howard called 'the

> Wheel of Life' — must turn continuously in place, so that *the law of return is kept and nothing is wasted.* For this to happen in the stewardship of humans, there must be a cultural cycle, in harmony with the fertility cycle, also continuously turning in place. The cultural cycle is *an unending conversation* between old people and young people, assuring the survival of local memory, which has, as long as it remains local, the greatest practical urgency and value .... The fertility cycle turns by the law of nature. The cultural cycle turns on affection (23).

Berry's concerns with the 'unending conversation' organize the work here. The Wheel of Life must turn in place. What happens when it does not? Fellow rural advocate, Wes Jackson (2011), describes an improper relationship with the industrial cycle[3] which grinds against the fertility cycle:

> Nature's ecosystems are ancient. They are real economies. The law of return described by Sir Albert Howard operates. They can be trusted. The materials sector, the industrial, is recent. It has no time-honored discipline to draw on. Skyscrapers, freeways, and suburbia have been made possible due to our discovery and use of fossil carbon, *not any organizing concept.* Soils and forests have fed and sheltered us but they too are in decline even with the subsidy of fossil fuels (15-16). [Emphasis added]

The lack of an organizing concept is the point. Do what humans make in relationship with creation preserve or deplete the integrity and means of the fertility and cultural cycles? What can we learn when we apply ourselves to the special discipline of learning about the real economies that we live in? Let us start by describing certain fundamental relationships and the consequences of getting those right or wrong. These relationships in Howard's 'Wheel of Life' require description and exploration, beginning with proper scale.

Right and Wrong Relationships, and Their Consequences

Proper

Proper understanding of proper scale starts by recognizing that we human beings are a *social species* and that our brains are big primarily to relate to one another. Psychiatrist and trauma specialist Bessel (2016) makes clear that we are wired for relationships and trauma disrupts this:

> Our culture teaches us to focus on personal uniqueness, but at a deeper level we barely exist as individual organisms. Our brains are built to help us function as members of a tribe. We are part of that tribe even when we are by ourselves, whether listening to music (that other people created), watching a basketball game on television (our own muscles tensing as the players run and jump), or preparing a spreadsheet for a sales meeting (anticipating the boss's reactions). Most of our energy is devoted to connecting with others (129).

Van der Kolk says, 'almost all mental suffering involves either trouble in creating workable and satisfying relationships or difficulties in regulating arousal.'[4] When we are estranged, we suffer. We know this. We torture individuals by isolating them or by having the 'tribe' turn against them. Proper scale begins at the tribe level.

Another problem with scale is with human relations and larger creation. Gustavo Esteva, the Mexican activist and former World Bank economist, reoriented the second half of his life to better understand the folly of large-scale human enterprises, grand schemes – well beyond families, tribes, and fertility cycles – to better warn others. Esteva and Prakash (1998) says, 'What is usually called "the Western project" attempts to universalize all of its institutions; that is, to impose everywhere the same set of behaviors, attitudes, arrangements, norms, and rules. The best model of this is the Catholic Church (31).' The imposition ruins, especially now that economies are globally integrated. Berry (2005) says, 'We identify arrogant ignorance by its willingness to work on too big a scale, and thus to put too much at risk. It fails to foresee bad consequences not only because some of the consequences of all acts are inherently unforeseeable, but also because the arrogantly ignorant often are blinded by money invested; they cannot afford to foresee bad consequences (54).' Exceed proper scale for humans and ecosystems and *'things fall apart, the center cannot hold.'*

The theologist Watts et al. (1977) considers this problem in philosophical and religious terms. Watts says, 'The mythologies of many cultures contain, in varying ways, the theme that man has fallen from grace and has had to replace it with technology (112).' Something essential has been lost, a sense of the real balance and order of creation, and so a hasty technical replacement is made. Watts illustrates this with the Tao Te Ching:

When the great Tao was lost,

there came (ideas of) humanity and justice.

When knowledge and cleverness arrived,

there came great deceptions.

When familial relations went out of harmony,

there came (ideas of) good parents and loyal children.

When the nation fell into disorder and misrule,

there came (ideas of) loyal ministers Watts,112-113). [99*c*]

Watts says, "The conscious control of life seems to involve us in ever more bewildering webs of complexity so that, despite their initial successes, technics, create more problems than they solve (ibid).' On a more human scale and much closer to home, my wife frequently said when our daughters were young, 'A family that has a lot of rules is not a family.' This meant to me that the rules were replacing active relationships responding to real needs and interests under live conditions. Proper scale and consideration enliven local agents.

The ruin from economics on too large a scale is everywhere in the US. Strip-mining, deforestation, soil loss, pollution of the air, water, and land, the industrial economy has made waste of many towns and rural places in the US and elsewhere. These are corpses piled up against the claimed benefits of this economy. I grew up in the Midwest in the middle of the 20th century and in many places once-thriving towns and villages have been abandoned and shuttered for allegedly failing to adapt up to the vagaries of the global market. It is difficult to gather and focus one's energies for local conditions when centripetal forces are so strong at the state, national, and global economic levels and when the industrial economy is the central assumption. Berry (2005), with the cultural cycle and commonwealth in mind, talks about it this way:

> We can't preserve neighborliness or charity or peaceability or an ecological consciousness, or anything else worth preserving, at the same time that we maintain an earth-destroying economy. Nothing ultimately flourishes in our present economy but selfish aims, and these are often mutually contradictory. We have to have a sort of pity for the CEO of a polluting corporation who desires wealth, healthy children, and a vacation in the restorative purity of nature. And surely we have to extend the same pity to those who are sure that 'it takes a village to raise a child' but who forget that it takes a local culture and a local economy to raise a village (77).

Here in the North Country of New York, the global industrial economy dictates living terms as well; and all around the state, the rural areas are on the losing end of the economic transactions with the five major cities and their suburbs. The transaction terms are clear: the rural places must send the cities produce, minerals, timber, and our children, and they return cheap products, prisoners, tourists, and their trash. Our children are warehoused away from community life and their generational place and are never asked to turn their eyes, ears, and consideration to understanding how where they live 'hangs together' or to becoming a fully qualified co-producer. Instead, they study abstract *data* according to state dictates, not local needs, allegedly to qualify them for exotic careers in an increasingly technical culture centered in the urban and suburban Meccas. Outside the schools, their imaginations are fed by pop culture formulas for consumptive lifestyles, always to be enacted elsewhere. They are lured away, Pied Piper-like, to better serve the economy, convinced as they are that they must brand and market themselves like any other product in the marketplace. But as Berry (1987) reminds us, our children are not only raised by the commercial culture to leave our communities, letting the fields go fallow and our communities collapse. Worse, they are raised to be our *enemies*, to become, in his words, 'itinerant, professional vandals,' ones who learn to treat every place as any place, valuable only for what can be taken from it (50). And these enemies are anxious, especially because of social media. But then again, they are taught by the educational institutions that they are valuable only for their productivity. It is about sales, not souls.

A culture that exports its children eventually dies, as does the Middle Earth relationship between people and the land. And when the Prodigal Son does not return – when Odysseus never makes it back to Ithaca – the promised redemption and reclamation never happens. The cycle stops.[5] In a mundane example, our local high school graduates here who launch themselves out of this town and into their careers, rarely return, and thus fail to reform, redeem, or supersede adolescent memories and sensibilities. They do not see their peers mature and their collective memories cannot be recollected and refined as a part of their own maturation. And family legacies fray with aging parents left behind.

The culture also protects less from the industrial, commercial economy. Arguably the best sign of a good culture is the health and well-being of its children. When I was growing up in the 1960s, commercial culture had limited time, range, and opportunity to get to me or to my peers. There were few commercial radio and television outlets, and they went off near midnight; and there remained an adult ethic that prohibited more than minimal marketing to kids. We stayed mostly in our neighborhoods, because we could meet most of our needs there and we were among trusted familiars. Then, I had the keen sense that the adults were still holding up the culture and that I could in turn take my time to grow.[6] Today, the marketing onslaught to kids is constant and families are brittler. Many homes have only one parent, and the common perception by children is that they are on their own and must make their way Hansel and Gretel-like with adults too busy, too disinterested, or too dangerous to protect, guide, or support. And instead of growing up fully credited and fortified to start their lives, American youth now often begin in debt. Rather than shielding them, the culture indentures them and forestalls their futures.

Proper Education

Public education too often abets the alienation from place and culture. The late English philosopher Sir Roger Scruton, in an interview (Peterson and Jordan 2018) calls it a form of 'de-bunking' education,

> ... always preferring de-bunking explanations of everything, reducing them to the lowest motive. "It's not truth but power that we pursue," and all of that Foulcauldian nonsense ... [T]he only response to that is to come up with 'bunking' explanations, so to speak, try to back in the subject matter one's own inherent belief in it, and to recognize that ... we're not around on this earth for very long and *we do have an obligation to find the things that we love and not the things that we reject*, and that those things that we love — the best way towards them is to look at the things that other people have loved. That's what a culture is. It's the residue of all of the things that people have thought worthwhile to preserve" (17:57-19:16).

Scruton continues:

> . . . The old way of teaching the humanities was . . . as objects of love. This is what I have loved, what previous generations have loved and handed onto me. Here: try it out, and you'll love it too. Whereas the postmodern curriculum is a curriculum of hatred. It's directed against our cultural inheritance. One after another, the works are paraded before us, stripped naked, and thrashed . . . by revealing whatever ideology or power structure is being concealed within them. And that of course is not why they were written and isn't how they should be understood . . . . And it means that the students who go through that curriculum come out with a sense of the hopelessness of everything" (1:11:05-1:14:11).

In a 'debunking' and otherwise explicitly economic education, the child does not become a qualified and sensible member of the culture but rather a vandal. And the study is indoors and abstract. She is taught to depreciate and take for her own. She is *mis*-educated, in the way that John Dewey meant it: misled. It is no wonder 'identity' is now a common, modern malady. Instead of inheriting and contemplating the value of her gifts, she is taught to reject them and is left with an unfurnished sense of self, self-defined instead of out of relationship. The Prodigal Daughter fails to return either physically or psychically.

By contrast, consider the traditional, enchanting forms of education by children in place described by Nabhan and Trimble (1994). It is developmentally essential, they argue, that children wander, find 'wonders,' identify them and with them, take them home, show them off, protect them, and imbue them with significance, often sanctity. No one has a possession like yours. It is unique, so you are unique. They give an example from the writer, Annie Dillard's book, Dillard (1994), when as a child she discovered a 1919 dime in her alley:

> Treasure was something you found in an alley. Treasure was something you dug up out of the dirt in a chaotic, half-forbidden, forsaken place far removed from the ordinary comings and goings of people who earned salaries in the light: under some rickety back stairs, near a falling-down pile of discarded lumber, with people yelling at you to get away from there . . . In spring I pried flat rocks from the damp streambed and captured red and black salamanders . . . In the fall, I walked to collect buckeyes from lawns. Buckeyes were wealth (4).

Nabhan and Trimble say:

> The key then is to plant the buckeye, feed the salamander, invent a story about the dime. With these acts of extension, *children begin to cultivate relationship* – and the concomitant risks and rewards of sharing, of giving, of love. By moving beyond simple ownership they avoid the trap of permanently linking their self-esteem which is only the first step – acquisition (26).

The authors described well the course of child development when the child is an active investigator and actor in his surroundings. Their work suits developmental psychologist, Alison Gopnik's arguments in, *The Gardener and the Carpenter* (2018). Against the modern cultural grain, Gopnik says that parents and teachers better serve children's development as *gardeners* rather than *carpenters*. Carpenters have explicit goals for how their child will turn out and plans (special classes, coaches, and camps) for how to get there. But adaptation to the ever-changing present and future has always been the sine qua non of the human species, Gopnik says. So, children will better adapt if nurtured, gardener-like, not directed so explicitly. Children learn mostly by watching and listening to adults and older children and they are geniuses at extracting the most important features of what they see and hear ('exploring to exploit'). Children should be wandering and wondering, Dillard-like, while adults protect and support. 'Yes, of course,' Gopnik would say to Roger Scruton, 'Share with children what you love. But don't try to force them to love or worship it as you have. They'll get it on their own and in ways you cannot imagine.' Add to wanderings a measure of what Robert Richardson (1986), the biographer of Henry David Thoreau, calls the *articulate observer* and you have a profound experience and existence under way.

> Aristotle and Pliny, together with Ruskin, recalled an increasingly scientific Thoreau to the importance, indeed the centrality and primacy, of the *articulate observer*. All could describe common sights and creatures so sharply as to make the description interesting — often exciting — after two thousand years. This continuing ability of the classical mind to comment on our own times was what had first attracted Thoreau to Virgil's agricultural writings (375). [emphasis added]

Great works of observation enrich our imaginations and join us to our ancestors. The ways of seeing and learning that Thoreau, Nabhan, Trimble, Dillard, and Gopnik describe also *preserve the means* of learning and curiosity, and enchant children with what and where they learn.

Proper Employment: Work

The industrial economy has long promised relief from the melancholy burden of work. Of course, this promise started much earlier. Once banished from the Garden of Eden, according to *Genesis*, we disgraced humans were forced to toil the soil for a meager existence. Work was the punishment for Original Sin. Oh, but technology, we are told, will soon reverse our sentence and return us to the garden of leisure to dwell with God. Christians and capitalists have the same heaven in mind: the end of work. But Wendell Berry (and many others[7]) question the validity – even the sanity of this workless dream life. In his poem 'Healing,' Berry (2010a) shows proper work as an active form of right relationship to serve creation and culture, the ways that work joins hands, neighbor to neighbor, past to present, young to old, creature to creation in a holy communion:

Good work finds the way between pride and despair.

It graces with health. It heals with grace.

It preserves the given so that it remains a gift

By it, we lose loneliness:

We clasp the hands of those who go before us, and the

Hands of those who come after us;

We enter the little circle of each other's arms,

And the larger circle of lovers whose hands are joined in a dance,

And the larger circle of all creatures, passing in and out of life,

who move also in a dance, to a music

so subtle and vast that no ear hears it except in fragments (10). [Emphasis added]

Rather than escape work, Berry suggests, we should ask what might be essential and irreplaceable in work. Union with creation is fair compensation.

A plumber friend shows this. Whenever he tears into old walls to do a repair, he always asks himself, 'What were the dead men thinking?' Extending this, the philosopher and craftsman Matthew Crawford (2009) says repair thankfully gets you out of your own head, out of your native narcissism, because to repair well you must imagine the machine's way of operating, from the machine's perspective, or from those who put it together. Crawford goes on to say, 'The satisfactions of manifesting oneself concretely in the world through manual competence have been known to make a man *quiet and easy*. They seem to relieve him of the felt need to offer chattering *interpretations* of himself to vindicate his worth. He can simply point: the building stands, the car now runs, the lights are on (15). Compare this to most modern work in a technically organized, hyper-fragmented institution; it is often hard to figure out if one's work is either worthwhile or even truly complete.[8]

For many years, I have been taking education students to tour and the local trade school, which is part of the public education system. I take them, in part, to see how different the learning environment is. Practically all of the trade students there are from the working class and their enthusiasm and contentment are obvious. There are real consequences to good and bad work and the students can experience them. Something is at stake. Poor work can mean a tree falls on a classmate or a plane from the sky. My students are astonished to hear a school counselor tell them that she has dealt with only five fights in 20 years and that the students will frequently show up at the school on holidays to keep working. Those are two good indicators of right human relationships and the value of good work. These teenagers are remarkably 'quiet and easy.'

So perhaps the end of work and the continuous pursuit of leisure is not heaven at all but its opposite: a hall of mirrors and a hell of self-doubt. Uselessness is dispiriting. Perhaps instead the goal should be not the escape from work and responsibility but the embrace of both. As the culture careens at increasing speed toward a worker-less economy, we must acknowledge the implications: what are humans for and what are the personal and social consequences of *chronically under- or unemployed humans*? And how can we preserve the means in right relationship without good work?

Identity and Work

Vexed identity is common today. On National Public Radio, it seems that every other story is about some individual's anxiety about identity and belonging along sex, gender, race, religious, or class lines. But once again, this is identity remarkably free of contexts, of place, time, and circumstance. Self-determined identity? To some, this is a preposterous notion. Vine Deloria (2003), First Nations philosopher, argued that identity questions largely affect *time people* and not *place people*. For *place* people, personality (identity) comes from place and character. For them, it is folly to imagine an identity devoid of place and circumstance. And as Bessel (2016) said, 'at a deeper level we barely exist as individual organisms' (129). These current obsessions with identity might indicate freedom and self-determination less than a spreading social disease: alienation resulting from a society in tatters. The social reality is reflected in many of the shows offered as entertainment. The long-running 'reality' show, *Survivor*, is a prime example of this dystopian trope. The rules are clear: you are on your own; you make the alliances you need to get to survive; human against human and God against all. Movie after TV show after video game poses the protagonist in a dystopia of social breakdown where only personal preferences and commitments matter. But these self-invented persons are utterly fragile and flattering urban fictions found in time cultures where the subsidies, penalties, and delusions are carefully kept off the books. They represent well the improper relationships between creation and human culture.

Conclusion

Right relationships are ones that preserve the means of human culture in tune with creation. They are measured by health and harmony and in gratitude. To live in proportion, on a human scale, increases integrity and sustainability, which reveals the insight that life is a gift to be appreciated and realized. Esteva and Prakash (1998) has advice for reducing global players to scale:

> Every transnational corporation, every imperialist government, every coalition of the Big and the Powerful, has an abstract, emblematic, illusory existence in the real world; they are but 'virtual realities' (non-realities). Coca Cola or Marlboro have no 'real existence' or power where people ignore them; they have no more power than the power people give them by 'believing' in what they offer. In most cases, people defeated colonial governments by turning against them the logic of their operations (transforming the benefit into a loss); doing so by dismantling the prevailing belief in their power ... The power of a colonial government ends at the very minute that the colonized cease to believe in it (31-32).

It is about what one believes in and ultimately commits to. On a more human level, there are the poetic voices from Wallace Stevens, Goethe, and Wendell Berry, who remind us of our power in the grace of our *humanness*:

In his poem, 'Chocoruna and its Neighbors,' Stevens (2015) says:

> To say more than human things with human voices,
> That cannot be, to say human things with more
> Than human voices, that, also, cannot be;
> To speak humanly from the height or from the depth
> Of human things, that is acutest speech (300).

Goethe (in Richardson 1986):

> To live within limits, to want one thing, or a very few things, very much and love them dearly, cling to them, survey them from every angle, become one with them — that is what makes the poet, the artist, *the human being*. And I shall never rest until I know that all my ideas are derived, not from hearsay or tradition, *but from my real living contact with the things themselves* (129). [Emphasis added]

Finally, and appropriately, Wendell Berry (1993):

> For one thing, though we have caused the earth to be seriously diseased, it is not yet without health. The earth we have before us now is still abounding and beautiful. We must learn again to see that present world for what it is. The health of nature is the primary ground of hope – if we can find the humility and wisdom to accept nature as our teacher. The pattern of land stewardship is set by nature. This is why we must have stable rural economies and communities; we must keep alive in every place the human knowledge of the nature of that place. Nature is the best farmer and forester, for she does not destroy the land in order to make it productive. And so in our wish to preserve our land, we are not without the necessary lessons, nor are we without instruction, in our cultural and religious tradition, necessary to learn those lessons (11).

By the end of the tour, students have walked and talked together and have started to consider at deeper levels what humans can and should do in shared spaces. Their reflections also show an increased appreciation of the past and desire to add to the commonwealth through the accumulation of knowledge, efforts, and commitment. This, in my view, is a good foundation for future public school teachers. If Neil Postman (1996) is correct, that a public school's job is to *create* a public (18), then we must get to it. But *here, in particular.*

Notes

1. Seen especially in the city building, designed by John Russell Pope, the architect of the Jefferson Memorial in Washington, DC.
2. The architectural firm from Florida never visited the site in upstate New York. The plans were approved in Albany, N.Y. by those who also never visited the site. The buildings were poorly insulated, poorly oriented, and flat roofed, in spite of annual heavy snowfalls.
3. William Blake describes the industrial cycle as 'wheels tyrannic.'
4. Van der Kolk's colleague, Peter Levine (2010), continues:'Our increasing use of drugs to treat [trauma] doesn't address the real issues: What are these patients trying to cope with? What are their internal or external resources? How do they calm themselves down? Do they have caring relationships with their bodies, and what do they do to cultivate a physical sense of power, vitality, and relaxation? Do they have dynamic interactions with other people? Who really knows them, loves them, and cares about them? Whom can they count on when they're scared, when their babies are ill, or when they are sick themselves? Are they members of a community, and do they play vital roles in the lives of the people around them? ... How can we help them feel in charge of their lives?' (526).
5. See Berry (2010b), 'The Work of Local Culture.'
6. Evolutionary psychologist, Alison Gopnik (2018), argues this extended adult protection of youth has been our key evolutionary adaptive advantage.
7. See also: Crawford (2016, 2009); Sennett (2008); and Langlands and Brockway (2019).
8. Crawford's motorcycle repair shop taught him 'the tangible elements of craft.' These were 'appealing as an antidote to vague feelings of unreality, diminished autonomy, and a fragmented sense of self that were especially acute among the professional classes.' (2009, 28).

Disclosure statement

No potential conflict of interest was reported by the author(s).

References

Berry, W. 1987. *Home Economics: Fourteen Essays*. San Francisco: North Point Press.

Berry, W. 1993. *Sex, Economy, Freedom & Community*. New York: Pantheon.

Berry, W. 2005. *The Way of Ignorance*. Berkeley, CA: Counterpoint.

Berry, W. 2010a. *Imagination in Place Essays*. 2010. Berkeley, CA: Counterpoint.

Berry, W. 2010b. *What Matters?: Economics for a Renewed Commonwealth*. Berkeley, CA: Counterpoint.

Bessel, V. D. K. 2016. *The Body Keeps the Score: Brain, Mind, and Body in the Healing of Trauma*. New York: Penguin.

Crawford, M. 2009. *Shop Class as Soul Craft: An Inquiry into the Value of Work*. New York: Penguin Press.

Crawford, M. 2016. *The World beyond Your Head: On Becoming an Individual in an Age of Distraction*. New York: Farrar: Straus and Giroux.

Deloria, V. 2003. *God Is Red: A Native View of Religion*. 3rd ed. Golden, Colo: Fulcrum.

Dillard, A. 1994. *An American Childhood*. New York: Harper Perennial.

Esteva, G., and M. S. Prakash. 1998. *Grassroots Post-Modernism: Remaking the Soil of Cultures*. London: Zed Books.

Gopnik, A. 2018. *The Gardener and the Carpenter: What the New Science of Child Development Tells Us about the Relationship between Parents and Children*. New York: Picador, Farrar, Straus and Giroux.

Jackson, W. 2011. *Consulting the Genius of the Place: An Ecological Approach to a New Agriculture*. Washington, DC: Counterpoint.

Langlands, A., and H. Brockway. 2019. *Cræft: An Inquiry into the Origins and True Meaning of Traditional Crafts*. New York: W.W. Norton.

Levine, P. 2010. *In an Unspoken Voice: How the Body Releases Trauma and Restores Goodness*. Berkeley, California: North Atlantic Books.

Nabhan, G. P., and S. Trimble. 1994. *The Geography of Childhood: Why Children Need Wild Places*. Boston: Beacon Press.

The Oxford English Dictionary Online, as of October 28, 2020, (https://oed.com/view/Entry/37261?redirectedFrom=commonwealth#eidOxford

Peterson, and Jordan, *Sir Roger Scruton/Jordan Peterson: Apprehending the Transcendent*, YouTube, December 14, 2018, video, https://www.youtube.com/watch?v=XvbtKAYdcZY

Postman, N. 1996. *The End of Education: Redefining the Value of Education*. New York: Vintage.

Richardson, R. 1986. *Henry Thoreau: A Life of the Mind*. Berkeley: University of California.

Sennett, R. 2008. *The Craftsman*. New Haven: Yale University.

Stevens, W. 2015. *The Collected Poems of Wallace Stevens*. New York: Vintage.

Watts, Alan, A. Chung-liang, and Huang. 1977. *Tao: The Watercourse Way*. New York: Pantheon.

Youth power – youth movements: Myth, activism, and democracy

Lynda Stone

ABSTRACT
This article explores relationships of youth power in a set of threads leading to the potential of today's youth activism to combat the climate crisis. Following an introduction featuring Sweden's Greta Thunberg, the threads are these: First from an American context is history of youth development, with one emphasis on the construction of adolescence. Second is learning experience about the US environment with its own national 'exceptionalist' history. Third is the role of inspiring youth movements, from history and contemporary times. Fourth is a turn to climate activism by youth as central to democracy. One general thematic about youth power across the article is that it oftentimes takes on a mythical character.

Introduction

The USA-based, *TIME* magazine's 'Person of the Year' for 2019 was Swedish teenager, Greta Thunberg, named for her leadership and impact on an international youth movement for climate change. Thunberg was the youngest person ever to receive the award in the nearly 100-year history of the magazine. The award is premised in 'the so-called Great Man theory of history, the notion that powerful individuals shape the world' (Felsenthat 2019, 49). Thunberg's award had its particular premise on the *Time* front cover in 'the power of youth.' While her teenage story might be seen initially as an incident of youthful rebellion, it became much more: a girl with Asperger Syndrome who skipped school to sit alone on the steps of Sweden's Parliament and who in her activist moment has inspired the world. Here is the dedication for the award:

> For sounding the alarm about humanity's predatory relationship with the only home we have, for bringing a fragmented world a voice that transcends backgrounds and borders, for showing all of us what it might look like when a new generation leads. (Felsenthat, ibid.)

Thunberg's story of inspiration briefly is returned to below. For this essay, a set of threads are brought together based on two interacting themes, the first is the 'emergence' of youth power and the second is its mythological presence. A first thread describes an American history of contributions by youth culminating in the emergence of a twentieth century 'youth culture' with its own mythology. A second thread connects the idea of youth to the nation's greater myth, that of American Exceptionalism and the place of geography, its natural environment, as central. A brief consideration of examples of environmental education is incorporated. A third thread turns to more specific activism in youth movements, one medieval clothed in myth and two in today's America. The exemplars are March For Our Lives and #BlackLivesMatter. A fourth thread turns to the commitment of the United States today toward combatting the climate crisis amid challenges to democracy. Here, the myth of youth power becomes an actuality.

Developing youth

A century's old myth is that children and youth are dependent on adults, their family and others, until they 'come of age.' In *The Rise and Fall of the American Teenager*, journalist Thomas Hine challenges this myth. A key theme of his account is that from America's beginnings to the end of the twentieth century, youth have been central to society's economy, their actual work significant for survival and family enhancement. He first describes systems of apprenticeship and indenture. Hine writes, 'Many parents had little choice but to depend on their children, either because of the exigencies of life on the frontier and the farm, or the wage scales of mines and factories forced them to' (Hine 1999, 121). Frontier and farm, it must be noted, serve as metaphors of life from settlement on, for immigrants across ethnicities and classes, in rural and urban life. Two historic trends lead into the twentieth century: Except for slavery, first, there existed a largely informal relationship of adults, children and societal struggle that changed into formal relationships outside of the family, especially with the growth of schooling. Second, in the name of reform, child labor laws and juvenile justice initiation changed relationships between youth and adults and by the middle of the twentieth century – to a different form of the power of the young.

Thus far, the history of American youth says little of which myth might exist, except perhaps in the sustainability of being young. Educational theorist specializing in youth studies, Nancy Lesko presents a view to complement Hine – a specific 'development' of youth–that itself is birthed in myth. Herein by the early twentieth century, relationships of adults and youth take on a 'darker' view, one in which adults might well be fearful of youth power. The myth begins in antiquity but becomes exemplified in the child psychology movement, especially by the 'father of adolescence', G. Stanley Hall. In Act Your Age! A Cultural

Construction of Adolescence, Lesko's theory links together two 'scientific' strains, the Great Chain of Being with recapitulation development and a romanticizing of youth.[1] What emerged is a belief accounting for youth power but retaining adult dependence, in progress coupled with a need for protection of adolescents, arguably through control. As a summary, Lesko explains that

> Hall idealized adolescence as the apex of human development—the period 'before the decline of the highest powers of the soul in maturity and age' and this romanticized view found widespread popular resonance .... Hall turned to the theory of evolution for a biologically based [ideal] of the special potential and pitfalls of adolescence .... [that led to suggestions for educational reform]. (Lesko 2012, 46-47)

The theoretical aim Lesko asserts is a society dedicated to perfection, including protection through youth. This analysis connects for her with hierarchical issues of nation, race, and gender. This hierarchy is understood as universal in an early twentieth century day but understood today as modeled on white, privileged, 'rural' males. A comment about other specific historic connections is youth movements promoting this mythical vision such as that of German Youth. Myths contribute to a lesser or better society.

Building from Lesko and returning to Hine, the idea of youth power was damped during World War I and II, with a brief, again mythical respite in the 1920s, a youth age of independence for some. Post-wars, the latter named as the 'age of the teenager' emerging in the 1950s, is illustrated first in film from white middle-class teens, Andy Hardy and Gidget.[2] In these middle century decades, what develops is a societal belief shared by young and old alike in youth power. In general, this remains today as a veneration of the young and being young. Beginning in the 60s and 70s, the consumer power of youth led and continues to lead American society in trends and values of style, music, film and the like. Youth work, by the way, continues across all class and racial lines. Poorer and working-class families, especially of minorities and immigrants, count on their children's contributions; and in middle-class families, the young work summer jobs, help pay for college, and especially enhance their own lifestyles. The emergence of youth culture as a world-wide late twentieth century phenomenon continues to develop. From Hine for the twenty-first century, youth will continue to hold power, a source for societal benefit that adults ought to support, not fear. Here is his closing statement: 'Rather than try to force young people to delay using their ... powers, we should accept youth as a gift .... In a rapidly aging society, youth may prove to be ... a source of energy to revivify a culture' (Hine, 304).[3]

Learning environment

The cultural value of youth, in being young, virginal and powerful across a millennial history becomes part of a belief in American Exceptionalism. As an emerging nation, this is tied to geography, of a vast continent to be settled, to turn a primitive nature into a civilized culture. Notwithstanding the exploitation of hundreds of sophisticated native cultures, use of abundant natural resources were 'for the taking.'[4] A mythical pristine New World was to be created separately from a worn out, decadent Old World. Belief in exceptionalism pervaded settlement incursions on all sides of the continent and then beyond: The City on the Hill, Manifest Destiny, the American Dream all combined across nineteenth century America's mythical belief in itself and in its youth.

In her very interesting study of exceptionalism, British historian Deborah L. Madsen offers several themes mythologizing America the nation, the frontier, and the West. To begin, from the literary contributions of mid-nineteenth century scholars such as Walt Whitman and Ralph Waldo Emerson, basic to a new country is a rugged individualism, itself arising from connection to nature. For example, Madsen writes this about Emerson:

> He looks, optimistically, to a future American culture that is self-reliant and complete in itself.... The fate of society is bound up with the fate of individuals.... Nature and not society in its present form will, in the future provide the true measure of personal value and achievement. Emerson is confident that nature would come to be the dominant force in American life, as the nation approaches its prescribed destiny .... [American] nature must become the model for an ideal American society. (Madsen 1998, 75)

By century's end the frontier is gone but often is eulogized for what it symbolized. The most famous account, found in every American history textbook, is the 'Turner Thesis.' In a lecture from the late nineteenth century, historian Frederick Jackson Turner 'defines the West not as a geographical place or region but as a process, a process that arises from and defines a unique American character .... [For Turner it grew out of] the experience of perennial rebirth, fluidity of social institutions, continual development and proximity to "the simplicity of primitive society"' (Madsen, 122, 123). Efforts from unique character result in completed continental conquest – with imperialist additions. In many ways domesticated nature in the form of settlements and indeed cities served and continues to serve culture. America's icon of this thesis, writes Madsen, is the Western film with its heroes and villains, its savagery masked as civilization and 'protection.'[5] Herein the 'unfettered individual' is tamed for the better, common good. Placed historically, with the last of the Indian Wars in the 1880s, settlement, i.e. conquest of the west, bled not only into the Gilded Age of great societal wealth but also into the Progressive Era's historical ambiguity. Both eras were beneficial for some but not for many others. Beyond the scope of this article but emblematic is the role of American President Theodore Roosevelt, in leading imperialist expansion with his 'rough riders' but also in contradictory early efforts in national land protection.

As part of 'patriotism,' American children and youth learn about the environment primarily through informal exposure and experience. This includes school trips and summer camps that are available to many if not most children. These typically are a regular feature of the middle school, grades 6–8 curriculum; many experiences also are organized by civic and religious organizations. For older more privileged youth such organizations as Outward Bound offer extended stays. In addition, there exist across communities, nature clubs and sports offerings. In the school curriculum, attention to nature is found in the elementary grades, appearing in traditional stories of frontier heroes and good deeds as part of reading and literacy development and in general attention to 'science.' By the time youth enter upper grades, the science curriculum becomes more discipline-based and environmental education largely is an 'elective.'[6] One place where brief environmental history is taught is in the social studies curriculum, significantly separate from science and the sciences.

Two historic accounts of learning environment are illustrative of educating youth in participatory origin and engagement. The first is the story of Smokey Bear, whose image first appeared in 1944. During WWII, Japanese warship shelling resulted in a serious fire in a California national forest. Government officials realized the general threat of fire that led to the creation of Smokey, who remains by 2014 'the third most recognized popular culture icon ... after Santa Claus and Mickey Mouse' (Wilson 2014, 121–122). In fact, in 1950 a badly burned bear cub from New Mexico was adopted as Smokey by the National Zoo in Washington. D.C. where he thrived until death in 1976. He is buried at Smokey Bear Historical Park in his native New Mexico. During the bear's life and the icon's history, children became junior rangers, beginning in 1964 have sent letters to the bear's own postal zip code, and with their families learned the motto, 'Only you can prevent forest fires.'

The second illustration is Earth Day, celebrated recently for its half-century history in a full-page, New York Times announcement: 'Fifty years ago, 20 million people took to the streets to demand that our government protect the health of our people and our planet. Out of that first Earth Day came bipartisan laws that keep us safe' (New York Times, February 2020, 30).[7] This event, occurring on 22 April 1970, had its own background that demonstrated a change from wilderness conservation and protection and public use to a broad ecological awareness. Several long-term trends, some pre-events and what became related causes led to the mass demonstrations and conservation activities of that day and its long-term effects: examples include the popularization of Rachel Carson's book, Silent Spring and Paul Ehrlich's The Population Bomb; trips into space and seeing earth from afar; activist civic efforts such as that of the Sierra Club; and the Free Speech, Anti-War and Back-to-the Land Movements. In an overview of the vital decade of the 1960s, Laban Carrick Hill includes Earth Day as significant for large-scale youth participation. He describes activities such as school-based planting of trees; in sum, he writes that the day became 'one of the most notable events in our nation's history' (Hill 2007, 138). In a longer study, The Genius of Earth Day, author Adam Rome writes this about the important efforts of youth:

> The most common images in media coverage of Earth Day were children picking up trash and protesting pollution .... In the suburbs especially ... [it] was a school-centered phenomenon .... If adults often seemed oblivious to the danger, kids were the great hope .... The greatest group of ... participants was K-12 students. (Rome 2013, 122)

Senator Gaylord Nelson of Wisconsin initiated the event but it might be said that its fifty-year history indicates that youth power has perpetuated the event, perhaps with the 'assistance' of their teachers.[8]

By the third decade of the twenty-first century, in America as well as other western developed nations, some environmental protection practices are taken for granted: people do pick up trash and littering is almost none existent, recycling is a public service, automobile owners pay attention to gas prices, and university ecology departments exist and scholars work with other disciplinary scholars to publish scientific evidence about the environment. As this article is being written, the Coronavirus pandemic reminds every person of the fragile relationship of nature and culture.

Inspiring movements

Before considering contemporary youth movements, another story clouded in medieval myth has been named 'the first popular crusade, the first ... youth movement' (Dickson 2008, xi). Perhaps too gone-by to be influential, this interesting story does provide inspiration and analytical insight. The story is told by American historian Gary Dickson, in his definitive modern mythistory, *The Children's Crusade* published in 2008. For him the historic event is a metaphor for the Sixties youth movement, so named 'for its predecessor' by the press. In pulling together anti-war, civil rights activists and others, college kids became grass-roots protesters for societal reform on campuses across the nation especially heading into the Democratic presidential convention in 1968.[9] In terms of response of the 'adult' world, Dickson quotes sociologist Todd Gitlin:

> There was enormous anxiety about whether the prevailing culture could hold the young .... It became easy to imagine that the whole of youth was regressing, or evolving into—what?
>
> Barbarism? A new society unto itself, a Woodstock Nation? A children's crusade. (Dickson 2008,7,Gitlin,T.(1987, 1993).*The sixties: Years of hope, days of rage. New York: Bantam.*)

In spite of trepidation, protest spread up and down the age ladder and across the society. By the end of the decade, there was federal legislation 'ensuring' civil and voting rights, including the desegregation of schools, that had to be implemented. Then, the War in Vietnam was ended and a corrupt president resigned. Today advances from the sixties and seventies that challenge 'exceptionalism' are still contested – and with backlash–there remains much to be done.

Returning back in time, in 1212 two separate swarms of French and German peasants formed an 'illegitimate' crusade to travel to Jerusalem to retrieve the True Cross. Not sanctioned by the Church yet 'inspired' by the crusading pope, Innocent III, and sparked by local events of 'divine inspiration,' thousands joined onto a large march, including many, many children. It was 'reported' that parents attempted to forestall youth participation but to no avail. It is also clear that from its inception and evolution, adults were present. Most of the hordes were not able to 'walk' across the sea to reach their goal and the crusade largely faltered. On its way, many died in imaginable and unimaginable ways. In the aftermath, many entered servitude in Europe or were made slaves if they did reach a distant shore. Some were merely assimilated into Italian villages. Given the generality of these claims, 'historians' might well have reprised what actually occurred; no one can actually know. What can be known is that myth and youth played a large role in its history.

Today, two contemporary youth movements in the United States offer inspiration for a possible third effort on climate change. These are March for Our Lives and #Black Lives Matter and each can be connected in broad strokes to the sixties protests with a very significant twenty-first century manifestation.

March for Our Lives (MFOL) is the activist movement initiated by Parkland Florida teenagers that has a direct link, her own inspiration, for Thunberg. Reported in the Time account, in May 2018, '[she] suggested … [to Swedish climate activists that] they emulate the … [the Parkland students] who had recently organized school strikes to protest gun violence in the U.S.' (Atler, Haynes, and Worland 2019, 58). Significantly they did not follow up but Thunberg did in her own Friday strikes.

As background to MFOL: The United States has a particular history of guns and gun violence. It does trace back to frontier America; it does trace to the Second Amendment to the US Constitution, legally transposed from the right to form a state militia into the right of individuals to possess firearms. Today a small percentage of the population own more guns than the nation's entire population: that is over 300 million! Politics and huge lobby funding support this ownership – with its own fear mythology.[10] Current highly politicized debates over police shootings/brutality/murder and are a part also.

Significantly, mass school shootings are endemic, in the popular imaginary initiated with the April 1999 rampage by two white adolescents at Columbine High School in Colorado. As reported in general media, 'School shootings' have occurred at all education levels, university to elementary school: Virginia Tech University, April 2007, 32 killed; Sandy Hook Elementary School in Connecticut, December 2012, 26

killed; Parkland, February 2018, 17 killed. Young people growing up in America today 'expect' shootings with locked facilities, the presence of armed resource officers, and 'duck and cover' drills daily reminders.[11]

For MFOL, the date etched in memories was 14 February 2018 at 2:21 in the afternoon at Marjory Stoneman Douglas High School in Parkland, Florida. An armed shooter entered what was known as the freshman building. Six minutes later 17 students and teachers were dead and 17 more were injured. The lives of everyone present, of the school and the town, were changed forever. The event became one of the worst mass shootings in American history – making the list above. In journalistic accounts of the day and its aftermath, characteristics of student responses stand out. An example is from Dave Cullen, author of Parkland: Birth of a Movement published in 2019. The book is a year-long travelogue with the Parkland Kids in which Cullen identifies a significant comparison: Unlike their predecessors of nearly twenty years, these youth turned their shock and grief from numbness to action. He sums his experience:

> I spent ten months shadowing these kids, and they were relentless, frequently racing around the country in opposite directions. That was their secret weapon: waging this battle on so many fronts with a host of different voices, perspectives and talents—healing each other as they fought. (Cullen 2019, 3)

Each student, of what became an early leadership cadre, decided to honor fallen comrades and their living peers who must face changed futures. They became political, established a national movement, and took up specifics of advancing gun control. In doing so, they encapsulated, captured, the recurring mythology – the power of youth. About a month after the shooting, their march on Washington, D.C., involved almost a million people and inspired hundreds of marches worldwide. It led to an extended bus tour and inspiration for specific legislation regulating gun ownership across America.[12]

The second inspiring movement is #BlackLivesMatter. It was 'born' in a July 2013 reaction, a Facebook post by Alicia Garza, to the acquittal of George Zimmerman for the shooting death of Tryvon Martin, a 17-year-old unarmed black youth a year before in Sanford, Florida. In his capacity as a neighborhood watch member, the armed Zimmerman claimed with ambiguous evidence to have been threatened by Martin. Jelani Cobb, staff writer for The New Yorker, reprises her post. Here is a part: 'I continue to be surprised at how little Black lives matter. And I will continue that. Stop giving up on black life .... Black people I love you. I love us. Our lives matter' (Cobb 2020, 19). Her friend, Patrisse Cullors,

amended the three words adding the hashtag and another, Opal Tometi, built a platform to spread the word. Importantly, it took a second killing, and in 2020 a third, for the movement to take off, gain momentum, and become a multi-racial, multi-generational movement. The second event occurred in summer 2014 with the death of Michael Brown in Ferguson, Missouri. The 18 year old was shot by a white police officer while walking in the street with a friend. In response, hundreds of protestors took to the Ferguson streets and remained there for many months. In intervening years, the movement maintained itself, then things dramatically changed with the agonizing death of George Floyd at the hands of police in Minneapolis, Minnesota on 25 May 2020. Caught on video, Floyd's death crystalized the issue of police brutality and killing especially of young black men, mobilizing a nation and hundreds of protest marches that occurred across the summer months. Going viral, there was also protest response internationally over similar issues.

As a movement #BLM has achieved its own identity, as Cobb reports in 2016, "'not your grandfather's civil rights movement' to distinguish its tactics and its philosophy from those of nineteen-sixties-style activism" (Cobb, 18). It remains distinct and credits its organizational form and activist activities to the Occupy Movement, a national moment that dates from a small protest encampment in a park in New York's Manhattan Wall Street financial district in September 2011. Its issue was wealth inequality between the infamous one and ninety-nine percenters, the great gap between the lives of the 'haves' and everyone else. Occupy was present in cities across the nation for many months and then fizzled out.[13]

Returning to #Black Lives Matter, the movement has a history that must be remembered. It is hundreds of years old beginning with the kidnapping of Africans sold into slavery and 'arriving' on the new continent more than 400 years ago.[14] It shares with MFOL thus a history of a nation's violence and shared goals for its significant change.[15] In The Making of Black Lives Matter: A Brief History of an Idea, published in 2017, Christopher J. Lebron writes this:

> [Blacks] have consistently found themselves at the business end of whites' chains, ropes, fists, guns, and nooses .... [This] is America's shame .... Slavery was an explicit practice of the openly legitimated such ownership and control. (Lebron 2017, 2)

There is much more: systemic racism is explicated in a series of needs for justice: health justice, economic justice, legal justice, cultural justice – the very 'idea' of full equality in their own country.[16] Two notes here: First for some American youth, climate justice is also incorporated in #BLM. Second, there is an age difference in the specific people and groups originating MFOL and #BLM whose work 'took off.' In their politics, the prior twentieth century distinction of adolescents and young adults seems to disappear.

Fostering democracy

Activism, the premise of this contribution, requires mass engagement via democracy. In the American context, it is time to come clean: In spite of many instances of youth engagement, the dangers of a climate crisis are not at the forefront overall of a committed culture – and thus the context seemingly differs from Europe. The United States is a highly polarized and currently politicized society that has complicated what a 'mass movement' can mean. Turning to this below, a first pass concerns obstacles to a movement: In recent years, there has been no national leadership on the crisis. Specifically, progressive proposals for a Green New Deal are attacked as 'socialist' and there is a widespread presence of science deniers. In addition, the prominence of other systemic issues either leave climate out or absorb it; these include public health and the pandemic, racism, and economic insecurity. Still, another problem is the vastness of the continent with climate issues that appear only to affect regions; thus, there is no 'national' problem. Consider what this says about the devastation of western wildfires, southeast hurricanes, and even mid-continent earthquakes resulting from energy fracking. The belief here is that natural events can be 'managed.' Finally, single political issues with specific aims in gun control and policing reform are treated piecemeal and responding to a more comprehensive, global condition seems impossible. This also means that small changes tend to have short lives through processes of advance and repeal.

To assert that there is no American mass movement on the climate crisis does not mean either its impossibility or the absence of ongoing endeavors to effect positive change. In this article, the two contemporary youth movements 'model' possibility. Connecting themes, they suggest that in the twenty-first century there has been a move from myth to actuality, that youth power may indeed be making a difference. Further, that a youth generation, while recognizing their own limitations, can 'move' their elders.

Before offering an example of American youth activism for climate reform, a brief return to Thunberg is instructive. Her personal school strikes did spark an international imagination. She delivered the remarks of a prescient teenager before such institutions as the United Nations, the World Economic Forum, the British Parliament – and her own in Sweden – and the US Congress. She has received international awards and been nominated twice for the Nobel Peace Prize.[17] An interview from October 2020 brings her life forward and exemplifies a central issue for youth movements: this is that youth do not remain young.[18] Across her experiences, Thunberg has solidified her belief in science, recognized the loss of empathy worldwide, enhanced support for countless activist efforts, and if she did not understand previously, used her platform for its time. Importantly, she has signaled her fulltime return, appropriate for a 17 year old, to being a student.

An American example of youth activism is found in the efforts of Jamie Margolin, a founder of Zero Hour with colleagues Nadia Nazar, Madelaine Tew, and Zanagee Artis in 2017.[19] Still, a high school student, she fits a large commitment to activism into the days of an average teenager. Among her accomplishments is publication of a guidebook for other youth activists.

A first step, she writes in Youth to Power: Your Voice and How to Use it, is to have a 'why,' a specific focus. Here is Margolin:

> My *why*, when you boil it down, is defending the sacred life of the Pacific Northwest. I could not live without the crisp air, the tall evergreens, the salty ocean, the birds, the salmon, the deer, the bears, the seals, the orcas, the smell of the cedars right after it rains. (Margolin 2020, 4)

In endeavors from activists like Thunberg, Margolin and countless others, there is a fostering of democracy beginning with an event, a vision, inspiration that cannot be denied. This particularity, however, exists with acknowledgement and support with others – for larger shared commitments to democracy and to justice.

Conclusion

This paper has offered a set of ongoing threads describing the emergence of the power of youth across American history. Youth, it has been posited, have always had a societal presence that across generations seemingly has become much more significant. A mythology has accompanied this importance: the young like the nation are exceptional. Importantly, with their power, today's youth have challenged an archaic conception of what counts as American Exceptionalism. America is a country that can survive and thrive attempting to reach its democratic ideal, especially through critique, through protest. The contemporary presence of two youth movements, March For Our Lives and #Black Lives Matter, provide a model for democracy, in an up-to-date conception of political activism. This is a conception that can emerge, already may be underway, with regard to a mass movement for combatting the climate crisis.

Histories of these mass movements have spurred their own proliferation and for fledgling efforts toward climate change. This is grassroots democracy taking hold, and largely through the internet. This will have to do without a national commitment. Ironically, two different forms of government need to co-exist today for a mass movement to take hold. At the grassroots level, decentralization means small groups organized around local problems, small democracies. In this leadership form, chapters like those in MFOL and #BLM may or may not have interconnection, some overall 'organization.' Importantly the model is that there is no hierarchy and no central unit 'ruling' the choice of action. At a national level, the aim of action differs; this is to legislate and implement large-scale policy. Typically, this entails the commitment and the power of a leader and/or control of a supportive a majority legislative party. The 2020 election that changed the presidency seems to signal a change with the placing of a priority on climate. Today, no matter who is in power, given a polarized society comprised in part by people who do not accept what science has revealed about climate, positive change will be exceedingly difficult to realize. Grassroots efforts in social movements, as those led by youth for their futures, do keep hope alive.

Notes

1. Recapitulation is the idea that the development of the young necessarily mirrors the development of civilization.
2. Missing in the teen films of this era is attention to racial and class diversity except for occasional tokenism.
3. Fear of youth power rises and falls within times, situations, and social hierarchies. Updating the text, some today fear of youth falls on minority youth, especially males of color. Moreover, delay of adulthood is the privilege of affluent youth, not all.
4. The desecration and destruction of indigenous cultures continues today.
5. While Madsen does not mention this, the image of white and black hats, good guys and bad, for Americans of a certain age requires no explanation! 'Protection' as a linking theme is an addition of the present author.
6. A contribution to youth learning is from personal communication, Associate Professor Janice L. Anderson, The University of North Carolina at Chapel Hill, USA.
7. Laws set up the US Environmental Protection Agency and regulations for the Occupational Safety and Health Act. Much of this effort has been repealed in the last few years.
8. As a young teacher in California, I taught on the first Earth Day and advised a middle school club that indeed did plant trees.
9. Dickson does not mention the college-based 'free speech movement,' of just a couple of years earlier. This was a protest to claim some control of university curriculum and governance.
10. The negative myth is this: 'They will take our guns. Looters and others will enter our neighborhoods and homes and harm our women and children.'
11. Duck and cover drills were actually practiced a generation or two previously to 'protect' against nuclear bomb threat. The same generalized fear has resulted again. As a reaction to an active shooter, school people are 'shelter in place.'
12. See from the Founders of March for Our Lives, Glimmer of Hope: How Tragedy Sparked a Movement available from https://marchforourlives.com/leadership/.
13. See Levitin, Michael Levitin M. (2015). 'The Triumph of Occupy Wall Street,' *The Atlantic*, https://www.theatlantic.com/politics/archive/2015/06/the-triumph-of-occupy-wall-street/395408/.
14. See the education project based on the inception of slavery in America, The 1619 Project, https://www.nytimes.com/interactive/2019/08/14/magazine/1619-america-slavery.html.
15. Support for #BLM appears on the MFOL website, https://marchforourlives.com/blm/.
16. See Tehama Lopez Bunyasi and Candis Watts Smith, Stay Woke: A People's Guide to Making All Black Lives Matter, published in 2019.
17. See Greta Thunberg from 2018, *No one is too small to make a difference*
18. See Oliver Whang, 'Greta Thunberg Reflects on Living Through Multiple Crises in a 'Post-Truth Society,' National Geographic, posted October 2020.
19. See Zero Hour at http://thisiszerohour.org/who-we-are/. A recent story of Margolin's life is by Brooke Jarvis in The New York Times Magazine, July 2020.

Disclosure statement

No potential conflict of interest was reported by the author.

References

Atler, C., S. Haynes, and J. Worland 2019. "The Conscience." *Time Magazine*, 50–55. December 21-30.

Bunyasi, T. L., and C. W. Smith. 2019. *Stay Woke: A People's Guide to Making All Black Lives Matter*. New York: New York University Press.

Cobb, J. 2020. "The Matter of Black Lives. The Political Scene." *The New Yorker*, 18–23. March 14, 2016 (reprinted July, 27, 2020).

Cullen, D. 2019. *Parkland: Birth of a Movement*. New York: Harper.

Dickson, G. 2008. *The Children's Crusade: Medieval History, Modern Mythistory*. New York: Palgrave Macmillan.

Felsenthat, E. 2019. "The Choice." *Time Magazine*, 48–49. December 21-30.

Founders of March For Our Lives, and C. W. Smith. 2018. *Glimmer of Hope: How Tragedy Sparked a Movement*. New York: Razorbill, Penguin .

Hill, L. C. 2007. *American Dreaming: How Youth Changed America in the Sixties*. New York: Little, Brown and Company.

Hine, T. 1999. *The Rise and Fall of the American Teenager*. New York: Perennial.

Jarvis, B. 2020. "Ecoanxious: A Teenage Climate Activist Grows up in the Shadow of an Uncertain Future." *The New York Times Magazine*, 28–33. July 26.

Lebron, C. J. 2017. *The Making of Black Lives Matter: A Brief History of an Idea*. New York: Oxford University Press.

Lesko, N. 2012 2001. *Act Your Age! A Cultural Construction of Dolescence*. 2nd ed. New York and London: Routledge.

Madsen, D. L. 1998. *American Exceptionalism*. Jackson, MS: University Press of Mississippi.

Margolin, J. 2020. *Youth to Power: Your Voice and How to Use It*. New York: Hachette.

Rome, A. 2013. *The Genius of Earth Day: How a 1970 Teach-in Unexpectedly Made the First Green Generation*. New York: Hill and Wang.

Thunberg, G. 2018. *No One Is Too Small to Make a Difference*. New York: Penguin.

Whang, O. 2020. "Greta Thunberg Reflects on Living through Multiple Crises in a 'Post-Truth Society." *National Geographic*, October 28. https//www.nationalgeographic.com/?url=https%3A%2F%2Fwww.nationalgeographic.com%2Fenvironment%2F2020%2F10%2Fgreta-thunberg-reflects-on-living-through-multiple-crises-post-truth-society%2F/

Wilson, R. K. 2014. *America's Public Lands: From Yellowstone to Smokey Bear and Beyond*. Lanham, MD: Rowan & Littlefield.

White, green futures

Cortland Gilliam

ABSTRACT

Black, Indigenous and otherwise minoritized communities of color are amongst the most vulnerable to the adverse consequences of environmental crises and the solutions proposed to remedy them. The participation and subsequent erasure of non-White youth activists and organizers within environmental sustainability struggles, and their subsequent erasure in global media coverage on climate activism has complicated any neat hierarchy of single concerns facing humanity. How is it that White and Western climate activists come to be the faces of the global youth climate movement? This essay examines the salience of race to the education and framing of environmental and sustainability efforts. It suggests that the separation of environmental concerns from the socio-political (and thus racial) contexts that produce them not only further marginalizes the contributions of non-White activists and the communities that they represent, but also engenders a violently White, and thus narrow, vision of a global, Green future.

Introduction

'... cropped ... purely on composition grounds' – David Ake (in Guardian, 25 January 2020)

The words above are those that the *Associated Press's* director of photography used to justify and excuse the removal of twenty-three-year-old, Ugandan climate activist Vanessa Nakate from a photograph featuring four other young climate activists – each of whom is White – in attendance of the 2020 World Economic Forum in Davos, Switzerland. In his explanation of photographer Markus Schreiber's decision to cut Nakate from the pre-published photo, Ake referenced Schreiber's belief that a building in the background of the photo, visible behind the Ugandan activist, was distracting (*Guardian*, 25 January 2020). The actual likelihood of the removal producing a more substantial and damaging distraction was effectively unthinkable to Schreiber at the moment of erasure. It is possible that the readers and interlocutors of the *Associated Press (AP)* – those who are both participants and observers in and of discourse, who constitute and signify education, the educator, and the educated – would have never noted this absence were it not for Vanessa Nakate's own alarm and her understanding of the imperative to sound it.

Upon realizing herself redacted from record, Nakate took to the digital counterpublic that is Twitter to speak/write her *self* back into fact, into history. In the almost eleven-minute video Twitter broadcast, 'What it means to be removed from a photo!' Nakate offers both prideful and pained reflections on her inclusion in the climate conferences at the World Economic Forum and subsequent exclusion from multiple news sources covering the event. In the viral video, the Ugandan climate activist confesses, 'This is the first time in my life that I understood the definition of the word racism' (@vanessa_vash). With this Nakate explicitly names the media's implication in the function of systems of racism and implicitly evokes the media's perpetuation of global anti-Blackness through its erasure of Black voices, stories, and legacies from sourcing and coverage. She speaks against and with what Michel-Rolph Trouillot might refer to as the first moment where silence enters into the production of history (1995). Her action exposes whiteness and its constitutive power relations, which undergird the representation of environmental and sustainability struggles writ large.

This indictment and its resultant pressure brought forth accountability from the *Associated Press*. After replacing the cropped photo with another featuring all five activists seated for the climate press conference – this time with Vanesa Nakate dead-center of frame – an *AP* spokesperson claimed that 'there was no ill intent' behind the removal, adding that the '*AP* has published a number of images of Vanessa Nakate' (*Buzzfeed*, 24 January 2020). However, as journalist Kenya Evelyn points out in their piece for the *Guardian*, 'The caption for the new image ... does not reference the switch or explain the previous cropping' (*Guardian* article). These details are important. By its own account, the *Associated Press* is an organization 'dedicated to factual reporting' and prides itself on being 'the most trusted source of fast, accurate, unbiased news in all formats' seen by 'more than half of the world's population ... every day'. The crisis of climate and the precarity of the natural environment, and consequentially of global ecology, exists in the present and regardless of intention. As Vanessa Nakate stated in her video broadcast, '[the crisis] ... it's already killing people' (@vanessa_vash). And a majority of the people who are dying *now*, and not solely in some distant future, are Black, Indigenous, and People of Color.

The stakes of this deliberate removal and its subsequent justification – one that did not engage with critiques based on an analysis of racism, colonialism, and capitalism – also cannot be ignored. This case, albeit anecdotal, becomes a prism through which to explore the problematics of representation and bias, of sourcing and documentation, and effectively, history and history making itself. The case of Vanessa's erasure represents that which is at stake in color-blind, race-neutral understandings of the global struggle for environmental sustainability, justice, and equity, and their attempted articulation via media. Vanessa Nakate's experience with the AP is not unique but indicative of much of the global media's failure to signal to the world with a self-awareness resultant from the internalization of the same investigatory gaze it applies to its objects of interest before reporting outward onto our timelines, newsstands, radios, and televisions (*Guardian*, 5 October 2019). As such, this case will be used to think race and power into our narratives around environmentalist movements and the futures they both pursue and forge.

Conceptual framing & central questions

From David Ake's defense of the erasure (in the epigraph), I derive the following building blocks for the analysis that will follow: *Crop. Pure. Composition. Ground.*

Crop – (1) the cultivated produce of the ground, while growing or when gathered; (2) to cut off the ends or a part of *Pure* – (1) free from anything of a different, inferior, or contaminating kind; free from extraneous matter; (2) being that and nothing else; mere *Composition* – (1) manner of being composed; structure; (2) the organization or grouping of the different parts of a work of art so as to achieve a unified whole. *Ground* – (1) the solid surface of the earth; firm or dry land (2) the foundation or basis on which a belief or action rests; reason or cause (Dictionary.com) With these conceptual units I ask: What does it mean for the presence and representation of racial and ethnic diversity – including its diversity of concern, of care, of knowledge, and of voice – to be considered antithetical or obstructive to a pre-determined and unified whole? What does the perceived imperative to crop – to cut off the ends and parts and to erase the extraneous – and, ultimately, cultivate a new vision in the name of structural integrity mean for one's vision of integrity, of structure, of what is deemed necessary and justifiable? What future crops are being harvested through such paradigms and practices of exclusion? What is the cost of alternatively structured visions with wider frames? More importantly, what is the cost of cultivating such a narrowed vision of and from the same over-tilled grounds of white supremacist reduction, negligence, and omission – the mere sociopolitical and cultural foundations and logics that have produced the crisis of climate and environmental precarity that imprisons all human and non-human life on this planet?

These are questions that this paper will examine and tease out – not to be answered in a generalizable, systematic, nor conclusive fashion, but engaged in an earnest attempt and invitation of reckoning. The failure to even approach a reckoning with these questions is to be resigned to violent indifference and to sustain a status quo of white supremacy, and its externalities of environmental miseducation and instability. In short, this paper explicates the salience of race in the framing, communication, and education of global environmental and sustainability struggle. It suggests that the separation of environmental concerns from the socio-political, and racialized, contexts that produce them both marginalizes the sustainability practices and contributions of non-White activists and the communities they represent, and engenders a narrow and violent understanding and vision of a healthy and safe environmental future.

The following review will explore processes and consequences of erasure for Black, Indigenous, and otherwise-marginalized persons within environmental and sustainability struggles through (1) the use of Critical Race Theory (CRT) as a framework for naming the exclusionary functions of whiteness within media and movement building, exemplified in the action of cropping; (2) a selective literature review of known environmental justice legacies and sustainability practices of Black and Indigenous communities, which historically ground the larger global environmentalist movements they are often cropped from; and (3) concluding reflections on the composition of 'green' futures where whiteness is primary, pervasive, and unnamed.

Review & analysis

Cropping for purity//'cropped . . . purely . . . '

The participation and subsequent erasure of non-White youth activists and organizers from around the world in environmental sustainability organizing, and their concomitant erasure from global media coverage on said organizing, has complicated any neat hierarchy of singular concerns facing humanity. Part of what the reality of such erasures represents, as in the case of Vanessa Nakate being cropped from the *AP* photo of climate activists at the World Economic Forum, is the salience of racial or racialized category and related notions of purity in the social imaginary of those who draw boundary, who decide what is within and outside of the frame, and who choose who is represented and who or what is given voice.

Important to understanding the regularity and taken-for-grantedness of such omissions is the idea that racism – particularly within the US context, but with resonance globally through various technologies of colonialism, capitalism, and anti- Blackness – is not aberrant, but rather endemic to and permanently imbedded within political and legal structures, and thus is foundational to political and social life (Bell 1991). This idea of the permanence of racism, also referred to as *racial realism*, is one of the core tenants of the Critical Race Theory (CRT) tradition, which originated in the mid-1970s as an offshoot of Critical Legal Studies (CLS). CLS scholars were influential in raising critiques against the dominant discourse within legal scholarship that held the instrument and institution of law to be objective, separate from politics and social control, and precise; instead, CLS emphasized the law's indeterminacy, incoherence, and its existence as a social contract that legitimizes the status quo (Bell 1991; Crenshaw 1988; Freeman 1977; Crenshaw et al. 1995).

However, the white-dominant CLS movement was unable or unwilling to extend its legal critiques to address the role of law in the legal-social construction of race itself (López 1997; Harris 1993; Crenshaw 1988). As a result, Critical Race Theory was conceptualized to name and analyze the ways that race, and thus racism, is constituted, codified, and reinforced in legal doctrine, custom, and jurisprudence (Crenshaw et al. 1995; DeCuir and Dixson 2004).

Because of the history and analysis that brought CRT to bear, it serves as a useful lens through which to understand racial marginalization and exclusion within and from narratives of environmental movements. Two constructs of CRT that I will focus on here are that of colorblindness (Bonilla-Silva 2014) and 'whiteness as property' (Harris 1993).

Color-blind racism

In the third chapter of *Racism Without Racists: Color-blind Racism and the Persistence of Racial Inequality in America*, Bonilla-Silva (2014) presents four central frames of what he calls 'color-blind racism': *abstract liberalism; naturalization; cultural racism; and minimization of racism*. For the purposes of this case, I am most interested in the frames of *abstract liberalism*, which uses "ideas associated with political liberalism ... and economic liberalism ... in an abstract manner to explain racial matters," and the frame of *naturalization*, which speaks to the ways that those positioned as White 'explain away racial phenomena by suggesting they are natural occurrences' (Bonilla-Silva 2014, 76).

Each of the frames of color-blind racism shares an attempt to mitigate impact by overemphasizing intention while simultaneously masking those very intentions as neutral. Regardless as to what intent was articulated after the fact, Markus Schreiber decided what he believed produced the best image. Schreiber's idea of a quality composition was dependent on the absence of a dark-skinned, Black, African woman.

This was a belief that was later co-signed by the director of photography, David Ake. When responding to a *Guardian* reporter about the issue, Ake's defense of Shreiber was that 'he thought the building in the background was distracting' (25 January 2020).

This defense is representative of color-blind racism's most fundamental frame of *abstract liberalism* specifically because of its emphasis on the liberal principles of choice and individualism (Bonilla-Silva 2014). In the justification, it was the photographer's right to choice that was defended. The defense of individual choice, however, neglects or obfuscates the uneven access that differently positioned individuals have to choice. If it was a matter of the building behind Nakate being distracting, rather than her presence, then she should have had the ability to choose to re-position herself. Although this scene is set in an international context, Bonilla-Silva would likely relate this defense of choice to the 'fallacy of racial pluralism – the false assumption that all racial groups have the same power in the American polity' (2014, 84). Additionally, the justification of choice in this instance requires that we ignore other potential and equitable choices that the photographer – the one privileged with relation to power – could have made at the moment of capture in order to achieve the desired absence of 'distraction'. Another way to view the *AP's* initial excuse of the erasure of Vanessa Nakate is through color-blind racism's frame of *naturalization*. Bonilla-Silva observes that this frame is used to "normalize events or actions that could otherwise be interpreted as racially motivated ... or racist" (2014, 85). Frequently, the frame of *naturalization* shows up as ironically race-neutral justifications for racial preferences for residential segregation, social groupings, or relationship pairings – preferences that may be viewed as harmless, natural, and possibly held by both marginal and dominant groups alike. In Nakate's case, we see it in the following statement to *BuzzFeed News* from an *AP* spokesperson: 'There was no ill intent. *AP* routinely publishes photos as they come in and when we received additional images from the field, we updated the story. *AP* has published a number of images of Vanessa Nakate' (24 January 2020). In this quote, the removal of Vanessa is wholly dismissed and relegated to 'routine' practice. According to the spokesperson, it is commonplace for the *Associated Press* to publish images and add additional images as they 'come in'.

Bonilla-Silva (2014) also acknowledges the many ways that these frames are often overlapping in their operation. We see such a combination with the first and most damning defense of the doctored photo, presented earlier in this paper: the photographer 'cropped it purely on composition grounds' (*Guardian*, 25 January 2020). Here *naturalization* is deployed through the unextraordinary deference to and confidence in the photographer's artistic license, professional taste, and expertise. It is to say that such a deference is to be expected. *Abstract liberalism* is deployed through the elevation of the photographer's artistic preference buttressed by an individualist paradigm that allows for and defends one's ability to select and deny on a case-by-case basis without consideration of group-based disadvantage.

Whiteness as property

The media's role in sustaining the status quo of white supremacy in its representations of environmental movements and their activists can also be seen through an understanding of 'whiteness as property'. Through study of the history of race in the United States, property scholarship, and the jurisprudence on race, Harris (1993) argues that "the concept of whiteness – established by centuries of custom (illegitimate custom, but custom nonetheless) and codified by law – may be understood as a property interest" (280). According to Harris, in creating property, which is to say the expectation of rights, "the inequalities that are produced and reproduced are not givens or inevitabilities; rather, they are conscious selections regarding the structuring of social relations" (1993, 280). Ultimately, Harris (1993) goes on to describe three property functions of whiteness afforded to those who are considered 'holders' of whiteness, those positioned as White: the exclusive rights of possession, of use, and of disposition (281). The attributes affiliated with these functions of property (within the liberal view of property) are (a) the right to alienability or to transfer; (b) the right to use and enjoyment; (c) and the absolute right to exclude (281).

The first function and related attribute of whiteness that is useful for this interrogation is that of possession, which includes the right to use and enjoyment. Harris (1993) notes that 'whiteness can move from being a passive characteristic as an aspect of identity to an active entity that – like other types of property – is used to fulfill the will and exercise power' (282). This aspect of whiteness shows up in Schreiber's decision to crop the original photo in that his decision was able to be cast as mere choice, making his and the other four climate activists' whiteness passive characteristics – which is to say making their whiteness a matter of sheer identity – while also utilizing their whiteness as a resource for image composition – a more active deployment of will and power. Additionally, the very ability for the photographer to benefit from a defense of his discriminating actions as neutral is effectively use and enjoyment of his whiteness. In this respect, Harris (1993) might argue that the right of dispossession and its attribute of transferability is also operative in the compounded enjoyment of whiteness by both the photographer and the director of photography.

The absolute right to exclude is one of the central principles of classical property theory (Harris 1993, 282), and perhaps the more readily apparent property function of whiteness relevant to Nakate's case. What was denied to Vanessa was her inclusion in the representation and documentation of youth climate activists at the conference, which also includes her exclusion from the benefits of said representation and documentation – benefits that were afforded to her White comrades. Additionally, she was even excluded from benefits that are sometimes afforded to those in proximity to whiteness. Finally, the defense of the composition of the 'purely' cropped image itself depends on the presupposed value attributed – whether consciously or otherwise – to white skin, and the contingent threat of black skin as contaminant. Through these property functions, the photograph became an extension of the photographer's, and the *AP's*, whiteness and thereby defensible without having to acknowledge race at all.

From the ground//'... on ... grounds'

While the sense of urgency around the state of the environment is felt by a countless many across the globe, it is not felt equally. Black, Indigenous and other minoritized communities are amongst the most vulnerable to the adverse consequences of present and mounting environmental crises and, in turn, the most vulnerable to any top-down global solutions proposed to remedy them. Additionally, there is a presumed futurity when discussing the threat of climate change and the inhabitability of the planet. These presumptions fly in the face of decades-long legacies of environmental activism and struggle by neglected, under-resourced, and marginalized populations for the protection of their local lands, neighborhoods, and foodways – all of which are actively threatened by capitalistic imperialism and industrial pollution in the now. Unfortunately, much of this history lives in silence when its representatives – adult and youth climate activists alike – are also excluded and erased from the mainstream narratives and spaces of environmentalism and sustainability solution-making. In her video broadcast that followed her discovery of being cropped, Vanessa Nakate cries, "We don't deserve this. Africa is the least emitter of carbons but we are the most affected by the climate crisis. But you erasing our voices won't change anything. You erasing our stories won't change anything" (@vanessa_vash).

Again, race and racism become central in understanding what is at stake in the global climate crisis and deterioration of the natural environment. They are the factors that help ground these larger global environmentalist movements in the lived experiences and knowledges of those most immediately and intimately impacted (Bullard 2001; Cole and Foster 2001; Ziser and Sze 2007). Ziser and Sze (2007) argue that the United States's unique position within and influence on global climate cultural politics makes consideration of these constructs particularly useful. In addition to being the world's greatest polluter and emitter of greenhouse gasses 'both in terms of its domestic industries and its financing and consumption of polluting industries elsewhere in the world', Ziser and Sze explain, the United States is also a great 'ideological force that drives much of current economic globalization' (2007, 385). This means that U.S.

history and its sociopolitical, cultural, and ideological products and practices – marketed to and disseminated throughout the world – bear consequence for global movements.

The role of race and racism in environmental movements within the US are well illuminated through the histories of (1) the Black-led environmental justice intervention into the dominant environmental preservationist movement, and (2) the traditions of indigenous community struggles for land and food sovereignty. The failure to recognize the racial implications of these legacies, which youth climate activists of color are often inhibited from representing on the global stage, is to reproduce environmentally destructive modalities and neglect insight into alternative ways of being in the world that emanate 'from the ground', on whatever grounds (Cole and Foster 2001). What follows will overview the environmental justice legacies of both Black and Indigenous communities and their relationship to mainstream movement building.

Environmental justice: form the natural to the ecological

Where do you pinpoint the beginning of a struggle for the right to a safe and healthy environment in a country such as the United States, with its centuries-long legacies of genocide, slavery, and capitalist exploitation? According to Cole and Foster (2001), in their foundational text *From the Ground Up: Environmental Racism and the Rise of the Environmental Justice Movement*,

... it more useful to think metaphorically of the movement as a river, fed over time by many tributaries. No one tributary made the river the force that it is today; indeed, it is difficult to point to the headwaters, since so many tributaries have nourished the movement (20).

Despite the many tributaries of this swelling movement, Cole and Foster argue that the environmental justice struggle finds much of its foundation in the Civil Rights Movement and the constituent grassroots networks therein (2001). Many of its early organizers were church officials or ministers who were already established within communities. As such, there is a tradition of local, community-based struggle as well as an intimate understanding of the significance of race and racism in group-based struggles for rights a sustainable living.

Following the Civil Rights Movement of the 1970s, the mass environmental justice movement as we know it today began in the predominately Black Warren Country, North Carolina in 1982, where more than 500 arrests were made during civil-rights-activist-organized protests of a PCB landfill (Bullard 2001). The protests inspired a US General Accounting Office study, which found that "three out four of the off-site, commercial hazardous waste landfills in Region 4 (which comprises eight states in the South) happen to be located in predominately African-American Communities, although African-Americans made up only 20% of the region's population" (Bullard 2001, 151). After breaking ground, the protests in Warren County inspired the United Church of Christ's Commission for Racial Justice's *Toxic Waste and Race* report, the first national study of its kind investigating correlation between waste facility sites and demographic characteristics (Bullard 2001; Cole and Foster 2001). This study found that race, more than land value, home ownership, or level of poverty, determined where waste facilitates were located (Bullard 2001).

In addition to providing a network of civil rights activists, with experience in direct action and grassroots movement building, the environmental justice movement made a major linguistic intervention into the mainstream environmentalist conversation by re-framing how we define and think about the 'environment'. The traditional environmentalist movement was primarily concerned with wilderness and wildlife preservation, resource conservationism, and, gradually, pollution prevention and cleanup (Ziser and Sze 2007; Cole and Foster 2001). The environmental justice movement widened the lens from the exclusively natural realm to the ecological. The environment at the focus of this intervention is the material environment "where we live, where we work, where we play, and where we learn" (Cole and Foster 2001, 16). The site of environmental hazard was expanded from the wild to the community, and, more specifically, to the poor community of color.

Due to the shift in framing, the environmental justice movement has been able to usher the term 'environmental racism' into popular discourse. Environmental racism is the descriptor given to the social, political, and economic forces that create disparate impact and distribution of environmental hazards on and in communities along racial lines (Cole and Foster 2001, 34). In Bullard's configuration, 'Environmental racism combines with public polices and industry practices to provide benefits for whites while shifting costs to people of color' and is "reinforced by government, legal, economic, political, and military institutions" (2001, 160–161). In short, one cannot separate the conditions that poor, predominately Black, Indigenous, and People of Color communities live in from the historical factors and decision-making apparatuses that produced and sustain those conditions – factors such as genocide, forced land removal, slavery, Jim crow, redlining, economic disinvestment, housing segregation, deteriorating housing, etc. (Bullard 2001, 153).

Sovereign struggle, strategies, and survival

The contemporary environmental justice movement also owes a great debt to the resistance legacies of Indigenous groups in the Americas, who, along with Indigenous communities around the globe, have been uniquely and vulnerably positioned in relation to the environmental hazard of climate change. Additionally, as with Black communities in the United States, Native American communities have been engaged in prolonged, local, and grassroots struggles for sovereignty over land and foodways (Bullard 2001; MacDonald et al. 2013; López et al. 2018; Cole and Foster 2001).

Furthermore, as with Black-led struggles, Indigenous peoples' contributions to environmental protection and education remain marginalized within media and within government-backed funding regimes.

Within the US context, one front of the environmental justice struggle being fought by Native American communities has been the resistance of land exploitation, which includes challenges to international industrial waste dumping, energy resource extraction, and water contamination. Native American nations are and have been prime targets for waste trading (LaDuke 1993; Bullard 2001). Additionally, due to institutional racism and disinvestment, many reservations and other Native American communities lack the infrastructure needed to deal with the health and economic costs that accompany some of these industrial processes. This is true of Indigenous populations not only in the US but also in poor, developing nations around the world that often become the sites of the waste from global capitalism. However, in the US, many small Native grassroot campaigns have been successful in blocking waste facility proposals through direct action, advocacy, and litigation (Bullard 2001; LaDuke 1993). Even so, some environmental changes have proved more difficult to overcome.

Control over the means and practices of obtaining food and sustaining community life have also been impacted by capitalist industry and climate control. In a study of Inuit youth of Rigolet, Nunatsiavut, Canada, researchers MacDonald et al. (2013) explore the ways that youth experience the impacts of climate change within the areas of changing travel conditions and access to hunting, challenges to Inuit culture, concern for Elder well-being, emotional and mental health consequences of altered practices, and potential adaptation strategies. Their study emphasizes the centrality of foodways not only to physical health but for the spiritual life of indigenous communities through its centrality to cultural practices, skill education, the sharing of traditional ways of knowing (López et al. 2018; MacDonald et al. 2013). The commonality of the shared threat to indigenous communities, their foodways, and their land, has led to global conferences where various indigenous groups have come together to address the impact of climate and environmental change, as well as the continued impact of colonial practices on indigenous lifeways.

Native American resistance traditions, too, have modelled how to fight for self-determination – the ability to dictate policy decisions in the interest of the community and speak for oneself – which has contributed to local, grassroots philosophy of the modern environmental justice movement (Cole and Foster 2001). Moreover, Indigenous resistance legacies speak to the comprehensive framework of analysis evoked by the term 'environmental justice'. The term environmental justice is intentional and is used, as Cole and Foster say, because it "requires democratic decision-making, community empowerment, and the incorporation of social structure – for example, existing community health problems, cumulative impacts of pre-existing environmental hazards, the effect of segregate housing patterns – in environmental decision-making processes (2001, 15–16).

All of the above is lost in the moment that a photographer decides that such a rich, race-conscious, anti-injustice, and grassroots legacy is a distraction. At the moment of the crop, the photographer erases the histories of intimate struggle with changes to climate and environment. He erases the knowledge and strategies for survival that those histories hold for the past and present. He loses the language for living in and with environment. He erases the ground on which he stands – on which the movement he aims to capture rests, and from which life springs.

Composing whiteness//'purely . . . composition . . . '

Not only did the cropping of Vanessa Nakate deliberately remove her presence, and incidentally the presence of the Ugandan community she represents, it also produced an entirely new image – one aligned with the artistic vision and journalistic judgement of the photographer. The produced image, which was sent out to *Associated Press* customers worldwide, was that of a White future. Perhaps, a future that the photographer viewed on some conscious and/or unconscious level to be one worth protecting. We must acknowledge what is at stake when a White career photojournalist does not consider the potential impact of his aesthetic preferences on the world. It is less about the potential for bad press than the impact and implications for the educative intent of his journalism, which is to say his ability to represent with accuracy, fidelity, and care. In her video following the discovery of the newly produced image, Nakate asks, "Does this mean that I have no value as an African Activist or that the people of Africa have no value at all?" (@vanessa_vash). Certainly, few media outlets would dare answer that question in the affirmative. As the president and CEO of the *Associated Press*, Garry Pruitt, said in an *AP* story on the incident and following apology, 'Our values are to cover the world – not the white world, but the whole world. And we need to do it' (27 January 2020). But we must ask, what do these values look like in practice?

Pruitt's statement reveals that whether the consequence was the desired, the action itself was not accidental. Moreover, the intention seemed to be to focus on Greta Thunberg – as has been the case in many outlets' coverage on recent youth climate activism. If anything, this trend further demonstrates how White faces have come to represent *the* face of the youth environmental movement, despite decades of non-White youth activism from around the world. Through so-called 'journalistic error' or artist discretion, whiteness becomes the frame of reference. This is true even when stories feature non-White activists, as in the example of 8-year-old climate activist Licypriya Kangujam, who has confronted media outlets for referring to her as the 'Greta of India' despite her activism existing before Greta reached global prominence (@ajplus, 27 January 2020). "If you call me 'Greta of India,' you are not covering my story. You are deleting a story" she told *AJ+* digital news. Therefore, when the four White, Western, and young women made the published *AP* photo, they were effectively being forwarded as the sole inheritors and saviors of the (White) global climate future.

Conclusion

The initial decision to eliminate Vanessa Nakate's presence at the climate conference via visual omission was not accidental but deliberate. Furthermore, it was deemed justifiable and commensurate with and within the existing practices of representation at the *Associated Press* organization. This can hold true even though the *AP's* Executive Director, Sally Buzzbee, issued the following *AP* statement later that day:

We regret publishing a photo this morning that cropped out Ugandan climate activist Vanessa Nakate, the only person of color in the photo. As a news organization, we care deeply about accurately representing the world that we cover. We train our journalists to be sensitive to issues of inclusion and omission. We have spoken internally with our journalists and we will learn from this error in judgment (24 January 2020).

How do we simultaneously hold this 'regret' and 'care' with the care and regret that inspired Nakate to speak back to her silencing? Moreover, how do we reconcile and hold ourselves accountable to a reality in which Shreiber was, in fact, committing himself to 'accurately representing the world' while enacting psychic, emotional, epistemic, and, ultimately, anti-Black violence on the Ugandan climate activist and woman?

The ultimate measure of a resolution to the climate crisis will not solely be found in the victory of a novel, universally applicable regulation – one antithetical to the undergirding logics of global capitalism that sustain the present degradation of the planet – but in the modification of our relational politics and constitutive policy towards vulnerable human and non-human life. As Cole and Foster (2001) point out, environmental justice is not solely about halting the construction of waste dumps and other energy facilities in under-resourced communities of color. It is also 'about creating clean jobs, building sustainable economy, guaranteeing safe and affordable housing, and achieving racial and social justice' (17). There is no way to accomplish the latter without addressing racial disparity at all levels of climate crises' inputs, impacts, and solutions. For this reason, it is paramount that Black and Indigenous and other non-White voices, like Nakate's, are represented in the unfolding composition of global public discourse and in the solution-building process on the climate. Put differently, any vision of a 'Green' future for the planet, by necessity, cannot be White.

Disclosure statement

No potential conflict of interest was reported by the author.

ORCID

Cortland Gilliam http://orcid.org/0000-0002-9442-4466

References

Bell, D. 1991. "Racial Realism." *Connecticut Law Review* 24: 363.

Bonilla-Silva, E. 2014. Racism without Racists: Color-blind Racism and the Persistence of Racial Inequality in the United States. MarylandNew York: Rowman & Littlefield Publishers

Bullard, R. D. 2001. "Environmental Justice in the 21st Century: Race Still Matters." *Phylon* 49 (3/4): 151–171.

Cole, L. W., and S. R. Foster. 2001. *From the Ground Up: Environmental Racism and the Rise of the Environmental Justice Movement*. Vol. 34. New York: NYU Press.

Crenshaw, K., N. Gotanda, G. Peller, and K. Thomas. 1995. "Introduction." In *Critical Race Theory: The Key Writings that Formed the Movement*, edited by K. Crenshaw, N. Gotanda, G. Peller, and K. Thomas. New York: New Press: xii-xxxii.

Crenshaw, K. W. 1988. "Race, Reform, and Retrenchment: Transformation and Legitimation in Antidiscrimination Law." *Harvard Law Review* 101 (7): 1331–1387.

DeCuir, J. T., and A. D. Dixson. 2004. ""So When It Comes Out, They Aren't that Surprised that It Is There": Using Critical Race Theory as a Tool of Analysis of Race and Racism in Education." *Educational Researcher* 33 (5): 26–31.

Freeman, A. D. 1977. "Legitimizing Racial Discrimination through Antidiscrimination Law: A Critical Review of Supreme Court Doctrine." *Minnesota Law Review* 62: 1049.

Harris, C. I. 1993. "Whiteness as Property." *Harvard Law Review* 106 (8): 1707–1791.

LaDuke, W. 1993. "Whitewashing Native Environmentalism: Reservation-Based Organizations and the National Pie." *News from Indian Country*VII, no 8. Ethnic News Watch Database. April 30: 1. http://libproxy.lib.unc.edu/login?url=https://www.proquest.com/docview/367628939?accountid=14244

López, B. M. J., D. Danielle, A. Patricia, and S. Soto. 2018. "Respecting Our Science and Ways of Knowing: Indigenous Peoples Food Sovereignty, Traditional Lifeways and Climate Change." Cultural Survival Quarterly 42, no. 4. Ethnic News Watch Database: 14. http://libproxy.lib.unc.edu/login?url=https://www.proquest.com/docview/2159620522?accountid=14244

López, I. H. 1997. *White by Law: The Legal Construction of Race*. Vol. 21. New York: NYU Press.

MacDonald, J., S. L. Petrasek, A. Harper, and V. L. Cunsolo Willox; Rigolet Inuit Community Government. 2013. "A Necessary Voice: Climate Change and Lived Experiences of Youth in Rigolet, Nunatsiavut, Canada." *Global Environmental Change* 23 (1): 360–371.

Trouillot, M.-R. 1995. *Silencing the Past: Power and the Production of History*. Boston: Beacon Press.

Ziser, M., and J. Sze. 2007. "Climate Change, Environmental Aesthetics, and Global Environmental Justice Cultural Studies." *Discourse* 29 (2): 384–410.

Spiritual exercises in times of climate change

Daniel P. Gibboney

ABSTRACT
'Facts remain robust only when ... supported by a common culture,' observes Bruno Latour. Current debates over the veracity of climate change are, in actuality, crises of facts. Questions of facticity have, moreover, precipitated a deeper issue – the prospects of unshared, 'alternative' worlds. Climate science believers have one world, climate change deniers another, creating what Latour calls 'epistemological delirium.' Following Latour, the paper turns to Pierre Hadot's description of Stoic physics and understanding of philosophy as *spiritual exercise*. Finally, taking up both Latour's claims of 'alternative worlds' and Hadot's notion of *spiritual exercise*, this paper explores the possibilities of shared practices in light of unshared realities.

Introduction

'You're entitled to your own opinion, but not you're not entitled to your own facts.'[1]

On 7 October 2020, Vice-President Mike Pence outed himself as an epistemologist. In a debate with then-Senator and Democratic Vice-Presidential Candidate Kamala Harris, Pence posited his own theory of knowledge. According to Trump's vice president, an individual is entitled to an opinion, but not one's own register facts. On its face, this seems self-evidently obvious. Facts are indisputable. How one interprets facts can be legitimate matters of disagreement. Furthermore, Pence's proposed liaison between fact and opinion implicitly entails a theory of political action. How one interprets facts will dictate the political posturing that one maintains. In this schema, facts represent both a general consensus and common ground for political life and contestation. The only problem with this position is that it is completely wrong. Vice President Pence would be completely correct if he were not categorically wrong. More accurately, one is not entitled to an opinion. An individual is, however, entitled to one's own facts.

Prior to the COVID-19 pandemic, entitlement to one's own facts was perhaps nowhere clearer than on the issue of climate change. In the United States, there is little to no consensus on either the facticity or legitimacy of the claim that the climate is changing and its mutation is caused by human activity. Furthermore, one's political position is not dictated by how one interprets the facts. Rather one's environmental posturing is dictated by the facts one accepts. Noticeably absent in this entire order is opinion. As such, an individual's position is dictated by the register of facts that one accepts. The action, or inaction, one advocates vis-à-vis climate change is not political; it is eminently factual.

This knotty arrangement of facticity, politics, and climate is taken up here. Any discussion of climate change and facticity, especially in the United States, invariably includes science and whether Americans 'believe the scientists.'[2] Accordingly, this paper focuses on the issue of climate change and the question of political action. In this effort, the essay at hand attempts three things. First, provide a reading of Bruno Latour's *Down to Earth*, specifically focusing on his notion of epistemological delirium. In *Down to Earth*, Latour suggests a characterization of contemporary American politics and climate change that centered on rejection of facts and disputation of science. Latour rightly identifies the failure of science to sufficiently defend itself and mobilize the public around the issue of climate change. Second, I submit for consideration Pierre Hadot's notion of philosophy as a *spiritual exercise*. Hadot's account of Stoic physics and its relationship to philosophy as a way of life provides an alternative historical arrangement of science and philosophical exercise than what is typically imagined. Third, in light of Latour and Hadot, I raise the prospects of contemporary spiritual exercises as forms of political activism around the issues of climate change.

***Epistemological Delirium,* and the failure of facts. Or, the politics of science**

Latour's 'hypothesis' in *Down to Earth* is simple. First, 'we can understand nothing about the politics of the last 50 years . . . [unless] the question of climate change and its denial [is] front and center' (Latour 2018, 2). Anything other than the opinion that humanity has collectively entered a 'New Climatic Regime,' it is impossible to 'understand the explosion of inequalities, the scope of deregulation, the critique of globalization, or, most importantly . . . the 'rise of populism' (ibid., 2) that dominate contemporary politics. On initial inspection, Latour's position seems straightforward. However, as in nearly all things, self-evidence is often deceiving.

In Latour's estimation, the realities of contemporary life are the outcome of an immensely complex, albeit 'single phenomenon.' Confronted with the reality of a changing planet, 'elites have been so thoroughly convinced that there would be no future life for everyone that they have decided *to get rid of all the burdens of solidarity as fast as possible*' (ibid., 18). Put slightly another way, the conditions of collective life have been precipitated by 'a single historical situation . . . the ruling classes' (known today rather too loosely as "the elites") . . .

conclu[sion] that the earth no longer had enough room for them and for everyone else' (ibid., 1).

Since the 1980s, abandoning even the rhetoric of a commonly inhabitable world (ibid., 9), 'the ruling classes stopped purporting to lead and began instead to shelter themselves from the world' (ibid., 1–2). Nevertheless, this pivot is a not retreat. Rather, shelter in Latour's usage involves at least two moves. First, a serious and concerted determination to flee from a mutually beneficial world. Second, construction of an alternative world exclusively invested with elite's own interests. Both moves, in Latour's estimation, equate to a forfeiture of a common world.

The loss of a common world is more than rhetorical – it is part and parcel of a new human condition – a 'wicked universality to be sure ... the new universality consists in feeling the ground is in the process of giving away' (ibid., 9). Confronted by this new universal condition, humans are tasked with 'discovering in common what land is inhabitable and with whom to share it' (ibid., 9). The discovery of mutually inhabitable land and negotiating with whom to share territory is what Latour calls the collectivist option. Simultaneously, humans are also presented with a sectarian choice. This second option, according to Latour, entails both ignoring climate change as an existential threat and simultaneously taking 'protection behind a wall [in order to] ... prolong the waking dream' (ibid, 9). Importantly this dream, rightly qualified as 'American', is one in which nine or ten billion humans will never benefit. 'Elites' have opted for option two.

This 'choice' by 'people – whom we can call the obscurantist elites from now on' (ibid.,19) is necessarily active. Descriptive on this point is Latour's use of 'obscurantist.' The act of obscuring is both intentional and orchestrated. Latour observes that in recognizing 'their dominance was threatened,' obscurantist elites 'dismantle(d) the ideology of a planet shared by all ... (however,) such an abandonment could under no circumstances be made public' (ibid., 21). On the surface, Latour's entire schematic can seem 'too much like a conspiracy theory' (ibid., 21). Nevertheless, 'the effects' Latour notes 'are quite visible ... the most edifying of these effects is the epistemological delirium that has taken hold of the public stage' (ibid., 21).

Latour explains that 'the public does not fully realize that the issue of climate-change denial organizes all politics at the present time' (ibid., 24). It is not simply that climate change organizes political life. Rather, it is that the denial of climate change chiefly arranges collective life of which politics is only one part. Practically, according to Latour, this phenomenon operates as 'denegation.' This renunciation goes beyond negation, however. As such, denial is both positive and, ultimately, constructive. It is at this juncture that Latour introduces *Down to Earth*'s central concept – epistemological delirium.

In its most straightforward sense, epistemological delirium refers to the socio-cultural conditions in which 'expression[s like] "alternative facts"' are able to gain mobilizing traction (ibid., 22). On this point, Latour advances two distinct, mutually dependent, points. Latour's first point involves the matter of how obscurantist elites convinced 'the public' of climate denial's veracity. A second point, predicated on the first, describes how the 'flight' by obscurantist elites from a commonly shared world is a scientific and epistemological affair.

At one level, concerted and systematic campaigns to deny climate change represents a 'cold' betrayal of the general public at the hands of an exclusivist elite (ibid., 22). This deceit operates at multiple interstices and exploits numerous previously existent dispositions. By way of example, Latour notes that 'ordinary people have a great tendency to be skeptical,' and this is a starting point for obscurantist elites' campaigns to deny climate change (ibid., 24). The American public's propensity for skepticism has been "incited 'by billions of dollars (have been) invested in disinformation' (ibid., 24). The consequences of this investment have dominated climate politics in the United States for nearly the last half-century.

Case in point for Latour's point is how 'journalists have seized on the idea that the populus has become attached to "alternative facts" to the point of forgetting all forms of rationality' (ibid., 22). Similarly, political pundits regularly accuse these 'good folks of complacency in their vision, [highlighting] the[ir] deplorable indifference to the very idea of truth . . . (and) a condemnable indifference to the facts' (ibid., 22). On the one hand, it is true large swaths of the American public have becoming increasingly hostility to the notion of facts. Still, in Latour's estimation, these explanations represent an interpretive failure. In neither case of journalists nor pundits, 'do they stress the reason why ordinary people have decided – and rightly so, in their case too – not to believe in anything any longer' (ibid., 24).

At the level of public epistemology, obscurantists elites operate from two basic understandings. The first understanding is that 'no attested knowledge can stand on its own' (ibid., 23). And secondly, 'facts remain robust only when they are supported by a common culture, by institutions that can be trusted' (ibid., 23). In the case of climate change, the common culture and institutions being attacked is science, and, the matter of how facts are circulated. Practiced in tandem, obscurantist elites have attacked the veracity of climatology and replaced it with an alternative register of facts. These efforts, more specifically, have involved a particular and systematic dismissal of scientific fact (ibid., 21).

The consequences of these systematic attempts to disprove scientific fact are numerous. Dismissal of scientific fact also represents a sacking of science's legitimacy more broadly. Importantly, according to Latour, this situation is almost entirely of scientists' own making (Latour and Chakrabarty 2019). Since positivism science has defended its practice on the basis of self-evidence. It follows that science is defensible on the basis of facts, and scientific facts are self-evidently defensible. Therefore, science is self-evidently defensible. Latour claims this is, and always has been, an epistemologically 'bad defense.' What is more, this justification has also proven insufficient politically (Latour and Chakrabarty 2019).

Questions around science, fact, and defensibility are neither novel nor unique to Latour. Latour's critique, if not indebted to, remarkably parallels Martin Heidegger's discussion of modern science. Heidegger notes modern science is commonly characterized 'in contradistinction to medieval science by saying that modern science starts from facts while the medieval started from general speculative propositions and concepts' (Heidegger 1977, 271). As such, distinctions between ancient and modern science 'cannot (simply) be established by saying, there concepts and principles, and here facts' (ibid., 271).

Heidegger contends 17th century philosopher-scientists 'understood that there are no mere facts, but that a fact is only what it is in light of the fundamental conception, and always depends upon how far that conception reaches' (ibid., 272). Alternatively, modern science 'thinks it can manage sufficiently with facts, or other and new facts, while concepts are merely expedients that one somehow needs but should not get too involved with, since that would be philosophy' (ibid., 272). As such, 'the comedy – or rather the tragedy – of science is that one thinks to overcome positivism through positivism' (ibid., 272). Whereas Heidegger claims 'this attitude prevails only where average and supplemental work is done' (ibid., 272), it seems that mediocrity dominates contemporary American science, at least in how it defends itself publicly.

In a podcasted conversation with University of Chicago historian Dipesh Chakrabarty, Latour claims the idea 'that facts are indisputably true' dominated scientific discourse from the rise of positivism until the turn of the twentieth century when traditional academic science was challenged by oil funded 'industrial' scientists (Latour and Chakrabarty 2019, 11:40–12:15). Despite overwhelming challenges to this defense of itself, Latour contends that facts are still offered to combat dismissal of facticity. By virtue of offering no defense of itself on any other basis other than the inherent self-evidence of facts, traditional climate science has been unable to offer any substantive defense itself from within its own discourse.

In light of this failure, Latour notes the futility of exasperation by scientists and their supporters 'that the people "believe in alternative facts," when they themselves live in an alternative world, a world in which climate mutation occurs, while it does not in the world of their opponents' (Latour 2018, 25). Practically, this can seem prohibitively obscure; perhaps as the abstract musings of a retired French winery heir. Nevertheless, Latour is on to something. The problem is exactly that 'there are now several worlds, several territories, and they are mutually incompatible' (ibid., 26). Acceptance or denial of climate change constructs alternative worlds, insurmountable impasses for confronting the threat climate change poses to the existence of the human species. As such, denying or accepting the 'facts' of climate change is profoundly cosmological. What one accepts or denies constructs the world in which one lives.

If one accepts Latour's characterization, it necessarily follows that contemporary environmental politics and science are nothing more than a morass. The question of political activism on the question of climate change, as such, 'is not a matter of learning how to repair cognitive deficiencies, but rather how to live in the same world ... face up to the same stakes' (ibid., 25). Arguments for anything less than a fundamental reappraisal of both the science and the politics of the environment fall victim to 'the habitual vice of epistemology, which consists in attributing to intellectual deficits something that is quite simply a deficit in shared practice' (ibid., 25).

Hadot reading stoic physics. Or, spiritual exercises as a way of life

In his inaugural lecture at the College de France, Pierre Hadot claims

> we must take account of all the concrete conditions in which [the ancients] wrote, all the constraints that weighed upon them: the framework of the school, the very nature of philosophia, literary genres, rhetorical rules dogmatic imperatives, and traditional modes of reasoning (Hadot 1995, 61).

Hadot's point concerns the relationship between a sufficiently historicist understanding of history of thought on the one hand and contemporary philosophical activity on the other.

Hadot further contends that 'a profound difference exists between the representations which the ancients made of *philosophia* and the representation which is usually made of philosophy today' (Hadot 2002, 2). Stating his point more clearly, Hadot claims that 'ancient philosophy had nothing in common with our contemporary philosophers, who imagine that philosophy consists ... in inventing a "new discourse" ... all the more original the more it is incomprehensible and artificial' (Hadot 1998, 73). Since the time of Socrates, 'the choice of a way of life has not been located at the end of the process of philosophical activity, like a kind of accessory or appendix' (Hadot 2002, 3). Rather, the choice of a way of life occupied a primary position.

'Primary' in Hadot's usage carries multiple meanings. First, primary is meant in the sense of 'at the beginning, in a complex interrelation with ... a certain way of living and of seeing the world, and voluntary decision itself' (ibid., 3). Accordingly, philosophy is an exercise and set of practices predicated on 'a concrete attitude and determinate lifestyle, which engages the whole of existence' (Hadot 1995, 83). For the schools of Greco-Roman antiquity 'the philosophical act is not merely situated at the cognitive level, but on that of the self and of being' (ibid., 83). On this point, primary assumes its second meaning in Hadot's usage – the preeminence of practice. In other words, Hadot notes that in the world of Greco-Roman antiquity, 'philosophy was a way of life' (ibid., 265).

Importantly, the choice of 'way of living' was 'never made in solitude' (Hadot 2002, 3). In antiquity, aspiring 'philosophers came to attend the classes in the school of his choice as a function of the way of life practiced there' (ibid., 98). As such, the 'philosophical school thus corresponds, above all, to the choice of a certain way of life and existential option, which demands from the individual a total change of lifestyle, a conversion of one's entire being, and ultimately a certain desire to be and live in a certain way' (ibid., 3).

The matter of an existential choice postures towards, perhaps, the central concept in Hadot's entire philosophical oeuvre – *spiritual exercises*. By spiritual exercises is 'mean[t] practices which could be physical, as in dietary regimes, or discursive, as in dialogue or meditation, or intuitive, as in contemplation, but which were all intended to effect a modification and a transformation in the subject who practiced them' (ibid., 6). Arnold Davidson, arguably Hadot's chief exegete internationally, in his introduction to *Philosophy as a Way of Life*, notes that 'spiritual exercises were exercises because they were practical, required effort and training, and were lived; they were spiritual because they involved the entire spirit, one's whole way of being. The art of living demanded philosophy was a lived exercise exhibited in every aspect of one's existence' (Hadot 1995, 21).

Whereas earlier historians of Greco-Roman thought privileged the dialogue as chief form of philosophical activity, Hadot flips the script. In a Hadot-ian historiography the philosophical dialogues of antiquity operate as a second-order justification for how one inhabits the world. It could appear that Hadot entirely dismisses the significance of the dialogue for philosophical activity. But Hadot notes that the existential option 'implies a certain vision of the world, and the task of philosophical discourse will there be to reveal and rationally justify this existential option, as well as this representation of the world' (Hadot 2002, 3). In this way Hadot echoes a Stoic sentiment when they 'said that philosophical discourse was not philosophy' in that 'one speak of philosophy as an ellipsis that has two poles – a pole of discourse and a pole of action, outer but also inner' (Hadot 2009, 110).

Philosophical dialogues in antiquity predominately covered three subjects – physics, ethics, and logic. On this point, 'it is perhaps necessary to have recourse to the distinction proposed by the Stoics ... the parts of philosophy – physics, ethics, and logic – were not, in fact, part of philosophy itself, but rather parts of philosophical *discourse*' (Hadot 1995, 266–67). For the Stoics, 'physics aimed to justify the choice of life we have just mentioned, and to make explicit the way of being-in-the-world which it implies ... .physics was not developed for its own sake but had an ethical finality' (Hadot 2002, 128). Knowledge of the physical world intersected with ethics at the level of establishing a grounding for one's way of practices of in the world.

Here it is important to lay out exactly what is meant by physics.

The Stoics understood physics in a dual sense. First, there is physics 'which applies itself to studying the structure of bodies, the movements, and the causes of natural phenomena' (Hadot 2006, 44). This is physics in the 'proper sense,' analyses of the physical world and nature. Importantly, Stoics' treatises on 'physics did not claim to be propounding a system of nature which was totally rigorous in all its details' (Hadot 2002, 208). Systematicity and exhaustivity were not ideals striven after; the purchase of physics lay elsewhere.

Physics was the discourse speaking of the gods, placing 'them in relation to the phenomena of nature' (Hadot 2006, 44). In this sense, physics was '"theological," use[d] the method of allegorical exegesis, and [was] therefore the noblest part of philosophy' (ibid., 44). Anchoring this observation is Hadot's particular framing of Stoic theology – an understanding that differs slightly from the contemporary academic discipline of theology. Hadot observes two primary meanings of theology: (1) description of the system implied in theogonies and religious texts; (2) 'allegorical exegesis of myths'; and, theories of first principles (ibid., 44). While seemingly disjointed, contemplation of nature singularly links all aspects of Stoic theology. Furthermore, framing theology in this way directly connects it with physics as a philosophical dialogue. Like theology, Stoic conceptualizations of physics and our contemporary scientific discipline while occupying the same moniker are obviously not identical enterprises. Nevertheless, physics both then and now are similarly cosmological – offering definitive accounts of the material world in which one lives. Importantly, however, 'in antiquity physics was a discourse and not ... an experimental practice' (Hadot 2006, 159). And as a discourse, Stoic physics offered a description of the world.

The Stoic world is 'one single living being ... in tune with itself and self-coherent ... Everything in the world is related to everything else, all is in all, and everything needs everything else' (Hadot 2002, 128–9). Most important for the relationship of ethics to physics is not the content of what the Stoic universe is, per se. Rather, how, according to Stoicism, 'the rationality of human action is based on the rationality of nature' (ibid., 128). Furthermore, Stoic physics was the basis for all attendant ethical position based on 'will for self-coherence' (ibid., 128). A desire for self-coherence is 'the basis of the Stoic choice, appear[ing] as a fundamental law within material reality, internal to each being and to the totality of beings' (ibid., 128). The primacy of the Stoic choice is inextricably grounded in the material reality of the world. Accordingly, the Stoic continually tried to be 'conscious of being in contact with the entire universe at every moment, for the entire universe is implied within each present moment' (ibid., 210). Accordingly, contemplation of the present moment is central to the relationship of physics to Stoic philosophic practice.

This question of time and the present moment is of supreme importance for both the Stoics and Hadot. The relationship of concentration on the present, its relationship to the cosmos, and their fundamental association in Stoic philosophy introduces numerous ideas central to Hadot's treatment of physics and spiritual exercises. One such idea is what Hadot terms 'cosmic consciousness.' By this is meant the consciousness that one is part of the cosmos, 'and the consequent dilation of our self throughout the infinity of universal nature' (Hadot 1995, 266). As such, the Stoic choice consists in 'in being able to say "Yes!" to the universe in its totality, and there want what happens to happen as it happens.' (Hadot 2002, 211). Presently, in this sense, means both identifying one's own subjective finitude in the infinite expanse of the universe and readily offering assent to the providence of the cosmos – 'existing within the cosmos, in unique reality of the cosmic event, was held to be infinitely precious' (Hadot 1995, 268). In this way, present is that of place and subjectivity.

Present, in Hadot's usage, also assumes a temporal meaning. According to Hadot, Stoics are advised 'to live *in the present*, letting (oneself) be neither troubled by the past, nor worried by the uncertainty of the future' (ibid., 268). As such, present here is eminently temporal, posturing neither towards past or future, but rather one's occupation of the contemporary world. Coupled together 'consent to fate and to the universe, renewed on the occasion of each event, is thus lived and practiced physics' (ibid., 211).

Conclusion. Or, climate activism as a spiritual exercise

Latour provides an account of modern climatology's failures. Hadot offers a description of physics as the basis for philosophic practice in Greco-Roman antiquity. Rather than representing a temporal impasse, the more than two-millennia separation between Latour and Hadot's phenomenon of interest can offer a productive space for thinking about the relationship of scientific facticity and socio-political action. This temporal and philosophical interspace offers potentially fertile grounds for thinking about how to re-orient any liaison between contemporary political praxis and scientific practice. Central at this juncture is the question of contemporary *spiritual exercises* and its relevance for contemporary activism around climate change.

Importantly, in both Hadot's and my own considerations, spiritual exercises are not only possible but necessary for confronting any contemporary moment (ibid., 284). Spiritual exercises do not correspond to particular philosophical systems, specific sets of material conditions, or exact temporal social realities (ibid., 282). In fact, spiritual exercises have been, and continue to be, practiced across a wide expanse of time, space, and political condition. Put another way, if one contends that the philosophical schools of antiquity were characterized by a choice or form of life,

> which is then justified *after the fact* by a given systematic construction ... then it is easy to understand how one can remain faithful to one's choice of a form of life without being obliged to adhere to the systematic construction which claims to found it (ibid., 282).

Nevertheless, the possibility of contemporary spiritual exercises is not inevitable. Rather, the possibility of spiritual exercises necessarily entails a particular relationship between thought and practice. More specifically, the idea of spiritual exercises 'is precisely that it is not a matter of purely rational consideration, but the putting in action of all kinds of means intended to act upon one's self' (ibid., 284). As such, an individual does not need to be convinced of the facticity of spiritual exercises to undertake them. One need only attempt them to practice them. Spiritual exercises create habitual practices. And their defense, or even proselytization, lies in their impact on the practitioner.

This matter of habit raises another issue on the matter of activism and climate change – that of the individual. On the matter of environmental activism, there is very little to nothing individuals can do in their personal lives that will have any impact on reversing climate change. Meaningful, substantive action on the issue of climate can only be undertaken at the level of nation-states and multinational corporations. If one wants to posit something an individual can do it is attempt to sway heads of state and Fortune 500 executives to act on the issue. In the case of the United States, such attempts have fallen woefully flat. Furthermore, from the American example, it could be claimed that the inability to sway political and economic elites to act on climate change represents a failure of socio-environmental movements and democratic political action. However, if one takes Latour even moderately seriously, such a diagnosis is hardly accurate.

The failure of the United States to take meaningful action around the issue of climate change is not political. At a very basic level, politics assumes a certain register of commonality and shared terrain on which engagement can take place. It is precisely this shared terrain that the United States lacks. As Latour has rightly pointed out, climate change believers and refuters live in fundamentally different worlds (Latour 2018, 25). While the problem could very well be political, the antidote is spiritual.

The failure of climate activism to effect any meaningful change in the United States is simultaneously political and scientific. As such, the solution needs to involve fundamental reappraisals of both politics and science. Nevertheless, it should be noted that more political action would most likely be futile. What's needed is cosmological activism as a form of spiritual exercise. It is one thing to speak of spiritual exercises in the abstraction or as an individual posturing. It is quite another to offer spiritual exercises as recompense to the potential extinction of the human species posed by climate change.

Latour claims 'the Achilles' heel' of any text taking up the issue of politics or activism is that 'the reader can justifiably ask, at the end: "All this is well and good. The hypothesis is attractive ... but what are we to do with it, practically speaking, and what does it change for me?"' (Latour 2018, 90). Like Latour, I do not want my 'essay to disappoint' (ibid., 91). However, this essay will not offer anything specifically programmatic. This article will, however, by way of Hadot, offer an example.

In his essay 'Study of Nature as a Spiritual Exercise' (2006), Hadot takes up the matter of scientific inquiry and ethics. Hadot writes that Aristotelian science 'demands that we not choose any end other than knowledge, and that we seek knowledge for itself, without any other utilitarian consideration' (ibid., 186). By defining science this way, Hadot contends, Aristotle forwarded objective knowledge as a value in and of itself. From this observation Hadot turns to the example of Jacques Monod – mid-twentieth century French Nobel laureate, biochemist, and philosopher of science. More specifically, Hadot draws attention to Monod's writings on Aristotelian science's 'ethics of objectivity' (ibid., 186).

On this theme, Hadot relays Monod at length. More specifically, Hadot quotes from the Monod's *Chance and Necessity: An Essay on the Natural Philosophy of Modern Biology* – '"to place the postulate of objectivity as the condition of true knowledge constitutes an ethical choice, and not a judgement of knowledge, since, according to the postulate of knowledge itself, there could be no true knowledge prior to this ... choice"' (ibid., 186). By way of explanation, Latour points out that,

> for Monod, such a choice posits an ideal that transcends the individual ... in this perspective scientific research is a 'spiritual exercise' to the highest degree, because, in Monod's words, it presupposes an 'ascetics of the mind,' or an effort at transcending oneself and mastering the passions (ibid., 186).

Central here, for both Latour and Monod, is what is termed 'ascetics of the mind.' As a spiritual exercise, science's '*askesis* of objectivity' entails an individual scientist to acting upon, and consequently transforming, themselves. And, in the process of doing so practically, the individual scientific practitioner practically renounces their own individual subjectivity. Renunciation of individuality takes place by virtue of scientific research being considered at the level of practice 'an end in and of itself' (ibid., 187). Practice for practice's sake.

Noticeably absent from Hadot's presentation are facts. The operative utility of Monodian science is not the discovery of facts. Furthermore, for Monod, scientific practice is not defended on the basis of factual self-evidence or socio-economic utility. Science as a spiritual exercise is defensible only on the basis of itself and the transformation of the individual practicing it. The defense is the defender.

Climate change as a fundamental threat to survival of the species has not sufficiently mobilized the American public to act. Defending science on the basis of facts has failed. Furthermore, American environmental activism has proven unable to bring about any kind of substantive change in policy. Practically and programmatically, it seems long past time to re-consider science, activism, and the relationship between the two. Nevertheless, undertaking such a project entails re-orienting an individual's practices of both. For example, rather than thinking of physics as a self-enclosed system, climate science could instead be offered as but one part of how an individual inhabits the world. In other words, offer climate science as a philosophical dialogue – part and parcel of a broader set of practices for living in the world. In this way both activism and scientific practice could be repurposed as a spiritual exercise. But thinking about climate activism in this way requires one to first say forget the political, this mission is spiritual.

Notes

1. Susan Page (2020). 'Read the full transcript of the vice-presidential debate between Mike Pence and Kamala Harris.' *USA Today.*
2. The phrase 'believing the scientists' is frequently used by former President Barack Obama when stumping for then-candidate Joe Biden during the 2020 Presidential Election.

Disclosure statement

No potential conflict of interest was reported by the author.

References

Hadot, P. 1995. *Philosophy as a Way of Life – Spiritual Exercises from Socrates to Foucault.* (ed./intro. A. I. Davidson) Translated by M. Chase. Malden: Blackwell Publishing.

Hadot, P. 1998. *The Inner Citadel – The Meditations of Marcus Aurelius.* Translated by M. Chase. Cambridge: Harvard University Press.

Hadot, P. 2002. *What Is Ancient Philosophy?* Translated by M. Chase. Cambridge: Belknap Press of Harvard University Press.

Hadot, P. 2006. *The Veil of Isis – An Essay on the History of the Idea of Nature.* Translated by M. Chase. Cambridge: Belknap Press of Harvard University Press.

Hadot, P. 2009. *The Present Alone Is Our Happiness – Conversations with Jeannie Carlier and Arnold I. Davidson.* Translated by M. Djaballah. Stanford: Stanford University Press.

Heidegger, M. 1977. *Basic Writings – From Being and Time (1927) to the Task of Thinking (1964).* Edited by D. F. Krell. London: Harper Press.

Latour, B. 2018. *Down to Earth – Politics in the New Climatic Regime.* Translated by C. Porter. Medford: Polity Press.

Latour, B., and D. Chakrabarty. 2019. "Bruno Latour and Dipesh Chakrabarty: Geopolitics and the 'Facts' of Climate Change." *WB202: The Critical Inquiry Podcast* from *Critical Inquiry,* April 11. htttps://critinq.files.wordpress.com/2019/04/latour_podcast.mp3

Page, S. 2020. "Read the Full Transcript of the Vice Presidential Debate between Mike Pence and Kamala Harris." *USA Today.*

Index